Math Review Handbook

Common Core Standards

Henry Gu

Author: Henry Gu
Editor: Christopher Gu

Copyright © 2015 by Henry Gu. All rights reserved.

ISBN-10: 1514818930   ISBN-13: 978-1514818930

In loving memory of Peter Gu,

    My father and mentor,

        Who inspired me to write this book.

Preface

Facing the introduction of the new math common core standards, our teachers and students need a clear picture about the new curriculum and its contents.

This book has included all the topics required in the common core standards and is organized by the nature of mathematics. It consists of five major subjects: Basic Algebra, Basic Geometry, Coordinate Geometry and Functions, Trigonometry, and Probability and Statistics.

Readers can benefit from its concise format and numerous examples.

This is an ideal review book as well as a convenient reference book.

Acknowledgement

Thanks to my family for all their love and support.

And thanks to you, the reader, for choosing this book as your study guide. I wish you the best of luck!

Contents

I. Basic Algebra

1. SETS, NUMBERS, AND OPERATIONS .... 1.
    1.1 Set and Notation .... 1.
    1.2 Operations with Sets .... 1.
    1.3 Numbers .... 1.
    1.4 Properties of Operations .... 2.
    1.5 Ratio, Proportion, and Percentage .... 2.

2. ALGEBRAIC EXPRESSIONS AND OPERATIONS .... 3.
    2.1 Operations of Polynomials .... 3.
    2.2 Factoring Polynomials .... 3.
    2.3 Rational Expressions and Operations .... 4.
    2.4 Radical Expressions and Operations .... 4.
    2.5 Exponents .... 5.
    2.6 Evaluation of Algebraic Expressions and Formulas .... 5.

3. EQUATIONS AND INEQUALITIES .... 6.
    3.1 Linear Equations .... 6.
    3.2 Linear Inequalities .... 6.
    3.3 Absolute Value Equations .... 6.
    3.4 Absolute Value Inequalities .... 7.
    3.5 Quadratic Equations .... 7.
    3.6 Quadratic Inequalities .... 8.
    3.7 Rational Equations .... 9.
    3.8 Rational Inequalities .... 10.
    3.9 Radical Equations .... 10.
    3.10 Linear System of Equations .... 11.
    3.11 Quadratic-Linear System .... 11.

4. SEQUENCE AND SERIES .... 11.
    4.1 Sequence .... 11.
    4.2 Recursive Definition .... 12.
    4.3 Arithmetic Sequences .... 12.
    4.4 Geometric Sequences .... 12.
    4.5 Series .... 13.
    4.6 Arithmetic Series .... 13.
    4.7 Geometric Series .... 14.
    4.8 Infinite Series .... 14.
    4.9 The Number e .... 14.

II. Basic Geometry

5. LOGIC .... 15.
    5.1   Negation: not, Symbol ~ .... 15.
    5.2   Conjunction: and, Symbol ^ .... 15.
    5.3   Disjunction: or, Symbol v .... 15.
    5.4   Conditional Statements .... 15.
    5.5   Biconditional Statements .... 15.

6. POSTULATES .... 15.
    6.1   Postulates of Equality .... 15.
    6.2   Postulates of Inequality .... 15.
    6.3   Partition Postulate .... 15.

7. DEFINITIONS .... 16.
    7.1   Angles .... 16.
    7.2   Midpoint .... 16.
    7.3   Bisector of a Segment .... 16.
    7.4   Bisector of an Angle .... 16.
    7.5   Perpendicular Lines .... 16.
    7.6   Parallel Lines .... 16.

8. THEOREMS .... 16.
    8.1   Congruent Angles .... 16.
    8.2   Properties of Perpendicular Lines .... 16.
    8.3   Properties of Parallel Lines .... 16.

9. TRIANGLES AND PROOFS .... 17.
    9.1   Interior Angles and Exterior Angles .... 17.
    9.2   Triangle Inequalities .... 17.
    9.3   Median, Altitude, and Angle Bisector: Concurrence ... 17.
    9.4   Right Triangle .... 18.
    9.5   Isosceles Triangle .... 18.
    9.6   Equilateral Triangle .... 18.
    9.7   Congruent Triangles and Proofs .... 19.
    9.8   Similar Triangles, Ratios and Proportions ... 20.

10. INDIRECT PROOF .... 21.

11. POLYGONS .... 22.
    11.1 Quadrilateral .... 22.
    11.2 To Prove a Parallelogram .... 22.
    11.3 Trapezoid .... 22.
    11.4 Polygons .... 22.

II. Basic Geometry

12. CIRCLE — 23.
- 12.1 Angles of a Circle — 23.
- 12.2 Segments of a Circle — 25.
- 12.3 Theorems — 25.
- 12.4 Common Tangents of Two Circles — 26.
- 12.5 Arc Length and Area of Sector — 26.
- 12.6 Inscribed Circle — 26.
- 12.7 Inscribed Polygon and Circumscribed Circle — 26.

13. CONSTRUCTIONS AND LOCI — 27.
- 13.1 Three Types of Constructions — 27.
- 13.2 Five Fundamental Loci — 28.
- 13.3 Compound Loci — 28.

14. TRANSFORMATIONS — 29.
- 14.1 Transformation Rules — 29.
- 14.2 Composition of Transformations — 30.
- 14.3 Symmetry — 31.
- 14.4 Rigid Motion and Orientation — 31.

15. SOLID GEOMETRY — 32.
- 15.1 To Determine a Plane — 32.
- 15.2 A Line Perpendicular to a Plane — 32.
- 15.3 Dihedral Angle — 32.
- 15.4 Perpendicular Planes — 32.
- 15.5 Parallel Planes — 33.

16. GEOMETRIC MEASUREMENTS — 33.
- 16.1 General Rules — 33.
- 16.2 Formulas for Measuring in Two Dimensions — 34.
- 16.3 Formulas for Measuring in Three Dimensions — 34.
- 16.4 Error in Measurement — 36.

III. Coordinate Geometry and Functions

17. COORDINATE GEOMETRY — **37.**
    17.1 Coordinate Plane — 37.
    17.2 Slope, Midpoint, Distance, and Centroid — 37.
    17.3 Coordinate Geometric Proofs — 37.

18. RELATIONS AND FUNCTIONS — **38.**
    18.1 Relations and Functions — 38.
    18.2 Increasing, Decreasing, and Constant Functions — 39.
    18.3 Piecewise Defined Functions — 39.
    18.4 Odd and Even Functions — 40.
    18.5 Maximum and Minimum — 40.
    18.6 Average Rate of Change — 40.
    18.7 Composition of Functions — 41.
    18.8 Inverse Functions — 41.
    18.9 Functions Under a Transformation — 42.

19. IMPORTANT FUNCTIONS AND RELATIONS — **43.**
    19.1 Direct Variation — 43.
    19.2 Inverse Variation (Function) — 43.
    19.3 Linear Function (First Degree) — 43.
    19.4 Greatest Integer Function — 45.
    19.5 Linear Inequalities — 45.
    19.6 Absolute Value Functions — 45.
    19.7 Quadratic Functions and Parabolas — 46.
    19.8 Quadratic Inequalities in Two Variables — 47.
    19.9 Equations of Circles (Relations, not Functions) — 48.
    19.10 Exponential Functions and Equations — 49.
    19.11 Logarithmic Functions and Equations — 52.
    19.12 Conic Sections — 54.
        (1) Ellipses — 54.
        (2) Hyperbolas — 54.
        (3) Parabolas — 54.
    19.13 Polynomial Expressions and Functions — 55.
        (1) Polynomial Formulas — 55.
        (2) Standard Form — 55.
        (3) Polynomial Division — 55.
        (4) The Roots of the Function — 56.
        (5) Remainder Theorem — 56.
        (6) Factor Theorem — 56.
        (7) End Behavior — 56.
        (8) Sketch the Polynomial Function — 57.
        (9) Rational Zeros Theorem — 57.

III. Coordinate Geometry and Functions

20. COMPLEX MUNBERS — **58.**
   20.1 Complex Numbers — 58.
   20.2 Addition of Complex Numbers — 58.
   20.3 Subtraction of Complex Numbers — 59.
   20.4 Multiplication of Complex Numbers — 59.
   20.5 Division of Complex Numbers — 60.
   20.6 The Midpoint of Two Complex Numbers — 60.
   20.7 The Distance Between Two Complex Numbers — 60.
   20.8 The Modulus of the Product of Two Complex Numbers — 60.
   20.9 The Polar Form of a Complex Number — 60.
   20.10 Multiplication and Division in Polar Form — 61.
   20.11 Powers of a Complex Number in Polar Form — 61.
   20.12 Roots of a Complex Number in Polar Form — 61.
   20.13 Revisit Quadratic Equations — 61.
   20.14 Fundamental Theorem of Algebra — 62.
   20.15 Linear Factorization Theorem — 62.

21. GRAPHIC SOLUTIONS OF LOCI AND SYSTEM OF EQUATIONS — **63.**
   21.1 Equations of Loci — 63.
   21.2 Graphic Solutions of System of Equations — 64.
   21.3 Graphic Solutions of System of Inequalities — 64.

IV. Trigonometry

22. TRIGONOMETRIC FUNCTIONS — 65.
    22.1 Degrees and Radians — 65.
    22.2 Coterminal Angles — 65.
    22.3 Arc Length — 65.
    22.4 Trigonometric Fuctions — 65.
    22.5 Pythagorean Triples — 66.
    22.6 Cofunctions — 66.
    22.7 Unit Circle — 66.
    22.8 The Sign of Trigonometric Functions — 66.
    22.9 Reference Angle — 67.
    22.10 Basic Applications — 67.

23. TRIGONOMETRIC GRAPHS — 68.
    23.1 Graphs of Trigonometric Functions — 68.
    23.2 Graphs of the Reciprocal Functions — 68.
    23.3 Amplitude, Period, and Frequency — 69.
    23.4 The Graph of $y = a\sin b(x - c) + d$ — 69.
    23.5 Inverse Trigonometric Functions — 70.

24. TRIGONOMETRIC IDENTITIES — 70.
    24.1 Trigonometric Identity Proofs — 70.
    24.2 Sum and Difference Formulas — 70.
    24.3 Double Angle Formulas — 71.
    24.4 Half Angle Formulas — 71.

25. TRIGONOMETRIC EQUATIONS — 71.

26. TRIGONOMETRIC APPLICATIONS — 72.
    26.1 Area Formula — 72.
    26.2 Law of Sines — 72.
    26.3 Law of Cosines — 72.

V. Probability and Statistics

27. PROBABILITY — **73.**
    27.1 Venn Diagram — 73.
    27.2 Counting Principle (2 or more activities) — 73.
    27.3 Permutation and Combination — 73.
    27.4 Probability — 74.
    27.5 Binomial Probability (Bernoulli Experiment) — 75.
    27.6 Binomial Expansions — 75.

28. STATISTICS (Univariate Data) — **76.**
    28.1 Common Methods of Collecting Data — 76.
    28.2 Analyze Data — 76.
    28.3 Data Distribution — 78.
    28.4 Statistical Inference — 81.
        (1) Confidence Interval — 81.
        (2) Significance Level — 83.

29. STATISTICS (Bivariate Data) — **84.**
    29.1 Exploring Categorical Data — 84.
    29.2 Exploring Numerical Data — 85.
        (1) Regression Modeling — 85.
        (2) Line of Best Fit (The Least-Squares Line) — 86.
        (3) Other Regressions — 86.

Appendix

A1. Graphing Calculator — **87.**

A2. Algebraic Formulas — **88.**

Index

I. Basic Algebra

1. SETS, NUMBERS, AND OPERATIONS

1.1 Set and Notation

A **set** is a collection of distinct elements.

Set Notation:
Finite Sets:
  e.g. $\{1, 2, 3, 4, 5\}$, $\{a, b, c\}$
Infinite Sets:
  e.g. $\{1, 2, 3, 4, 5 \cdots\}$,
  $\{\frac{1}{2}, \frac{1}{4}, \frac{1}{8}, \frac{1}{16} \cdots\}$
Empty Set or Null Set:
  $\{\ \}$ or $\emptyset$
  e.g. $\{0\}$ is not an empty set.

Set Builder Notation:
  e.g. $\{x \mid 0 \le x \le 10,$ where x is a real number $\}$

Interval Notation:
  e.g. $(2, 5)$ represents $\{x \mid 2 < x < 5\}$
  $[2, 5]$ represents $\{x \mid 2 \le x \le 5\}$
  $(2, 5]$ represents $\{x \mid 2 < x \le 5\}$
  $[2, 5)$ represents $\{x \mid 2 \le x < 5\}$

e.g. Which of the following notation is equivalent to the set $\{1, 2, 3, 4\}$?
(1). $\{x \mid 1 < x < 4,$ where x is a whole number $\}$
(2). $\{x \mid 0 < x < 4,$ where x is a whole number $\}$
(3). $\{x \mid 1 < x \le 4,$ where x is a whole number $\}$
(4). $\{x \mid 0 < x \le 4,$ where x is a whole number $\}$
Here (1) is $\{2, 3\}$;
   (2) is $\{1, 2, 3\}$;
   (3) is $\{2, 3, 4\}$;
   (4) is $\{1, 2, 3, 4\}$
Answer: (4)

1.2 Operations with Sets

Universe or Universal Set:
The set of all elements under consideration.

Subset:
A subset is a set which is a part of a larger set.
  e.g. $\{1, 2, 3\}$ is a subset of $\{1, 2, 3, 4, 5\}$

Complement of a set:
e.g. Universal Set U = $\{1, 2, 3, 4, 5, 6, 7, 8\}$
   Set A = $\{1, 2, 3\}$
The complement of set A is denoted by $\overline{A}$ or A'.
   $\overline{A} = \{4, 5, 6, 7, 8\}$
The complement of set A has all elements in the universal set except the elements in set A.

Intersection of two sets: symbol $\cap$
The set of all elements that belong to both sets.

Union of two sets: symbol $\cup$
The set of all elements in either set.

e.g. A = $\{1, 2, 3, 4, 5\}$ and B = $\{2, 4, 6, 8, 10\}$
  then A $\cap$ B = $\{2, 4\}$
    A $\cup$ B = $\{1, 2, 3, 4, 5, 6, 8, 10\}$

1.3 Numbers

Counting Numbers or Natural Numbers:
    1, 2, 3, 4, 5, ...
Whole Numbers: 0, 1, 2, 3, 4, 5, ...
Integers: ..., -3, -2, -1, 0, 1, 2, 3, ...
Consecutive Integers: n, n+1, n+2, ...
  e.g. -3, -2, -1 ; 4, 5, 6
Consecutive Even Integers: n, n+2, n+4, ...
  e.g. -8, -6, -4 ; 0, 2, 4 (zero is an even integer)
Consecutive Odd Integers: n, n+2, n+4, ...
  e.g. -5, -3, -1 ; 7, 9, 11
Perfect Squares: 4, 9, 16, 25, 36, 49, 64, 81, ...
Prime Numbers: 2, 3, 5, 7, 11, 13, 17, 19, 23, ...
Rational Numbers can be written as an integer, a quotient of two integers, a terminating or repeating decimal:
  e.g. 2, -5, 1.25, 0.333..., 2.345345...,
  $\sqrt{4} = 2$, $\frac{4}{5}$, $\sqrt{\frac{4}{9}} = \frac{2}{3}$

Irrational Numbers are decimal numbers that neither repeat nor terminate:
  $\sqrt{3}$, 1.41421... , π
Real Numbers: All the above numbers.
Rounding: 3.456 rounded to the nearest integer is 3, to the rearest tenth is 3.5, and to the nearest hundredth is 3.46
Absolute Value of a Number: $|5| = 5$, $|-5| = 5$,
  $|12| - |-5| = 12 - 5 = 7$

e.g. The integers are a subset of the rational numbers.
   The rational numbers are a subset of the real numbers.
   The union of the rational numbers and the irrational numbers is the set of real numbers.

I. Basic Algebra

1.4 Properties of Operations

Commutative: $a + b = b + a$, $ab = ba$

e.g. $3 + 5 = 5 + 3$, $3 \cdot 5 = 5 \cdot 3$

Associative: $a + (b + c) = (a + b) + c$
$a \cdot (b \cdot c) = (a \cdot b) \cdot c$

e.g. $3 + (5 + 7) = (3 + 5) + 7$
$3(5 \cdot 7) = (3 \cdot 5)7$

Distributive: $a(b + c) = ab + ac$

e.g. $3(5 + 7) = 3 \cdot 5 + 3 \cdot 7$

Additive Identity: 0
$x + 0 = x$, $0 + x = x$

e.g. $5 + 0 = 5$, $0 + 5 = 5$

Multiplicative Identity: 1
$x \cdot 1 = x$, $1 \cdot x = x$

e.g $3 \cdot 1 = 3$, $1 \cdot 3 = 3$

Additive Inverse: $-x$
$x + (-x) = 0$

e.g. $-5 + 5 = 0$

Multiplicative Inverse: $\dfrac{1}{x}$

$x \cdot \dfrac{1}{x} = 1$

e.g. $3 \cdot \dfrac{1}{3} = 1$

e.g. If $\odot$ is a binary operation defined by
$a \odot b = a^2 + b^2$, then $3 \odot 4 = 3^2 + 4^2 = 25$.

1.5 Ratio, Proportion, and Percentage

If two ratios are equal, they are in proportion.
$$\frac{a}{b} = \frac{c}{d} \quad \text{or} \quad a \cdot d = b \cdot c$$
In a proportion, the product of the means is equal to the product of the extremes.

e.g.   The ratio of the three interior angles of a triangle is $2 : 3 : 4$. Find the measure of the largest angle.
$2x + 3x + 4x = 180$
$9x = 180$
$x = 20$
The measure of the largest angle is
$4x = 80$

Percent:

$\% = \dfrac{1}{100}$, $100\% = 1$

$60\% = \dfrac{60}{100} = 0.6$, $6\% = \dfrac{6}{100} = 0.06$

$0.6 = 0.6 \times 100\% = 60\%$
6% of $\$50 = 0.06 \cdot \$50 = \$3$

Percent of Increase or Decrease:

$$\frac{|\text{New Amount - Original Amount}|}{\text{Original Amount}} \cdot 100\%$$

e.g. The gas price increased to $\$2.50$/gal from $\$2.00$/gal.

$$\text{Percent of Increase} = \frac{|2.50 - 2.00|}{2.00} \cdot 100\% = 25\%$$

Tax Problems:

Tax = Base Price • Tax Rate

e.g. The tax rate in NYC is 8.5%. How much do you pay for merchandise tagged $\$50$?

Tax = $50 \cdot 8.5\% = 50 \cdot 0.085 = \$4.25$
Total Amount = $50 + 4.25 = \$54.25$

I. Basic Algebra

2. ALGEBRAIC EXPRESSIONS AND OPERATIONS

2.1 Operations of Polynomials

Addition and Subtraction

Combine Like Terms:

e.g. $3x^2 + x - 8 - x^2 + 5x + 4$
$= (3x^2 - x^2) + (x + 5x) + (-8 + 4)$
$= 2x^2 + 6x - 4$

e.g. $(3a + 5b - 7) - (2a - 4b + 8)$
$= 3a + 5b - 7 - 2a + 4b - 8$
$= (3a - 2a) + (5b + 4b) + (-7 - 8)$
$= a + 9b - 15$

Multiplication and Division

Use Distributive Property:

e.g. $3y(2x^2 + 2y^2 - 2)$
$= 3y \cdot 2x^2 + 3y \cdot 2y^2 + 3y(-2)$
$= 6x^2y + 6y^3 - 6y$

e.g. $(a + b)(a + b)$
$= a \cdot a + a \cdot b + b \cdot a + b \cdot b$
$= a^2 + 2ab + b^2$

e.g. $\dfrac{12x^4 - 4x^2}{4x^2}$ (same as multiply $\dfrac{1}{4x^2}$)
$= \dfrac{12x^4}{4x^2} - \dfrac{4x^2}{4x^2}$
$= 3x^2 - 1$

Formulas:

$(a + b)^2 = a^2 + 2ab + b^2$
$(a - b)^2 = a^2 - 2ab + b^2$
$(a + b)(a - b) = a^2 - b^2$

2.2 Factoring Polynomials

Factoring is the reverse process of multiplication.

Factor Out the Greatest Common Factor (GCF):

e.g. $3x^2 + 6x = 3x(x + 2)$

e.g. $18x^2y^3 + 12x^3y^2 - 6x^2y^2 = 6x^2y^2(3y + 2x - 1)$

The Difference of Two Squares:

$$a^2 - b^2 = (a + b)(a - b)$$

e.g. $4y^2 - 25 = (2y + 5)(2y - 5)$

e.g. $x^4 - y^4 = (x^2 + y^2)(x^2 - y^2)$
$= (x^2 + y^2)(x + y)(x - y)$

Trinomial:

Product-Sum Method by Trial and Error:

e.g. $x^2 + 2x - 15 = (x + a)(x + b)$

Try to find a and b such that
$a \cdot b = -15$ (Last)
$a + b = 2$ (Middle)
We have $5 \cdot (-3) = -15$ and $5 + (-3) = 2$,
therefore $x^2 + 2x - 15 = (x + 5)(x - 3)$

e.g. $2x^2 - 7x - 15 = (2x + 3)(x - 5)$

$2x \diagdown 3$
$x \diagup -5$

Here $2x \cdot x = 2x^2$ First•First = First
$3 \cdot (-5) = -15$ Last•Last = Last
$2x \cdot (-5) + x \cdot 3 = -7x$ Cross-multiply = Middle

Four-term Polynomial:

If the product of the first term and the last term is equal to the product of the second term and the third term, then the polynomial can be grouped.

e.g. $3x^3 - 6x^2 + 2x - 4$
$= 3x^2(x - 2) + 2(x - 2)$ grouping
$= (3x^2 + 2)(x - 2)$

Factor Completely:

e.g. $2x^3 - 14x^2 + 20x$
$= 2x(x^2 - 7x + 10)$
$= 2x(x - 2)(x - 5)$

I. Basic Algebra

2.3 Rational Expressions and Operations

Denominator cannot be zero.

e.g. $\dfrac{x}{x-2}$ when $x = 2$, it is undefined

e.g. $\dfrac{2x+1}{x(x^2-25)} = \dfrac{2x+1}{x(x+5)(x-5)}$
$x \neq 0, x \neq -5, x \neq 5$

Simplify: $\dfrac{3x^4y^4(x^2-25)}{2x^3y^5(x-5)} = \dfrac{3x^4y^4(x+5)(x-5)}{2x^3y^5(x-5)}$

$= \dfrac{3x(x+5)}{2y} = \dfrac{3x^2+15x}{2y}$

Multiply: $\dfrac{2x^3}{x^2-x-12} \cdot \dfrac{x^2-16}{6x}$

$= \dfrac{2x^3(x+4)(x-4)}{(x+3)(x-4) \cdot 6x}$ factor the numerator

$= \dfrac{x^2(x+4)}{3(x+3)}$ and the denominator first; cancel out common factors in the numerator and the denominator

Divide: $\dfrac{\dfrac{x-3}{2x+1}}{\dfrac{2x-6}{2x}}$

$= \dfrac{x-3}{2x+1} \cdot \dfrac{2x}{2x-6}$ multiply the inverse

$= \dfrac{(x-3) \cdot 2x}{(2x+1) \cdot 2(x-3)}$

$= \dfrac{x}{2x+1}$

Combine: $\dfrac{1}{x+1} + \dfrac{x-1}{2}$ LCD $= 2(x+1)$

$= \dfrac{2 \cdot 1}{2(x+1)} + \dfrac{(x-1) \cdot (x+1)}{2(x+1)}$

$= \dfrac{2 + x^2 - 1}{2(x+1)}$

$= \dfrac{x^2+1}{2(x+1)}$

2.4 Radical Expressions and Operations

$\sqrt{a \cdot b} = \sqrt{a} \cdot \sqrt{b}$ $a \geq 0, b \geq 0$

$\sqrt{\dfrac{a}{b}} = \dfrac{\sqrt{a}}{\sqrt{b}}$ $a \geq 0, b > 0$

Simplify: e.g. $\sqrt{75} = \sqrt{25 \cdot 3} = 5\sqrt{3}$

e.g. $\sqrt{\dfrac{9a^4b^6}{16}} = \dfrac{3a^2b^3}{4}$

Combine: e.g. $\sqrt{18x} - 4\sqrt{2x} = 3\sqrt{2x} - 4\sqrt{2x} = -\sqrt{2x}$

e.g. $2ab\sqrt{2b} + 5a\sqrt{18b^3} - 7b\sqrt{8a^2b}$
$= 2ab\sqrt{2b} + 5a \cdot 3b\sqrt{2b} - 7b \cdot 2a\sqrt{2b}$
$= (2ab + 15ab - 14ab)\sqrt{2b}$
$= 3ab\sqrt{2b}$

Multiply: e.g. $3\sqrt{2} \cdot 7\sqrt{2} = 3 \cdot 7\sqrt{2 \cdot 2} = 21 \cdot 2 = 42$

e.g. $\sqrt{y}((x - \sqrt{y}) = x\sqrt{y} - y$

Divide: e.g. $\dfrac{4\sqrt{6}}{2\sqrt{3}} = \dfrac{4}{2}\sqrt{\dfrac{6}{3}} = 2\sqrt{2}$

e.g. $\dfrac{\sqrt{98x^3y^5}}{\sqrt{2xy}} = \sqrt{\dfrac{98x^3y^5}{2xy}}$

$= \sqrt{49x^2y^4} = 7xy^2$

Rationalize: e.g. $\dfrac{1}{\sqrt{3}} = \dfrac{1}{\sqrt{3}} \cdot \dfrac{\sqrt{3}}{\sqrt{3}} = \dfrac{\sqrt{3}}{3}$

e.g. $\dfrac{\sqrt{5}}{3 - \sqrt{5}}$

$= \dfrac{\sqrt{5}(3 + \sqrt{5})}{(3 - \sqrt{5})(3 + \sqrt{5})}$ multiply the conjugate

$= \dfrac{3\sqrt{5} + 5}{4}$ of the denominator

$a + \sqrt{b}$ and $a - \sqrt{b}$ are **conjugates** of each other.
$(a + \sqrt{b})(a - \sqrt{b}) = a^2 - b$

I. Basic Algebra

2.5 Exponents

$a^0 = 1 \quad (a \neq 0)$ $\qquad 5^0 = 1, \quad (-5)^0 = 1$

$x^{-n} = \dfrac{1}{x^n} \quad (x \neq 0)$ $\qquad 5^{-2} = \dfrac{1}{5^2} = \dfrac{1}{25}$

$x^a \cdot x^b = x^{a+b}$ $\qquad 5^2 \cdot 5^3 = 5^{2+3} = 5^5$

$\dfrac{x^a}{x^b} = x^{a-b}$ $\qquad \dfrac{y^3}{y} = y^{3-1} = y^2$

$(x^a)^b = x^{a \cdot b}$ $\qquad (5^2)^3 = 5^{2 \cdot 3} = 5^6$

e.g. $(-5)^2 = (-5)(-5) = 25$
$\quad -5^2 = -(5^2) = -25$

e.g. $2x^2y^3 \cdot 5x^3y^3$
$= 2 \cdot 5 x^{2+3} y^{3+3}$
$= 10 x^5 y^6$

Rational Exponents

$x^{\frac{a}{b}} = \sqrt[b]{x^a} \qquad x^{\frac{a}{b}} = (\sqrt[b]{x})^a \qquad x \geq 0$

e.g. $x^{\frac{1}{2}} = \sqrt{x}, \quad x^{\frac{2}{3}} = \sqrt[3]{x^2}$

e.g. $125^{\frac{2}{3}} = (\sqrt[3]{125})^2 = 5^2 = 25$

e.g. $\dfrac{3^{\frac{1}{3}}}{3^{-\frac{2}{3}}} = 3^{\frac{1}{3} - (-\frac{2}{3})} = 3^1 = 3$

e.g. $\sqrt[3]{8x^{16}y^{10}}$
$= 2^{\frac{3}{3}} x^{\frac{16}{3}} y^{\frac{10}{3}} = 2^1 x^{5\frac{1}{3}} y^{3\frac{1}{3}}$
$= 2x^5 y^3 x^{\frac{1}{3}} y^{\frac{1}{3}} = 2x^5 y^3 \sqrt[3]{xy}$

Rational exponents are useful to simplify complicated expressions.

Scientific Notation:

$\qquad a \times 10^n \qquad 1 \leq a < 10, \ n$ is an integer

e.g. $23{,}000 = 2.3 \times 10^4$
$\quad 0.0043 = 4.3 \times 10^{-3}$

e.g. $(2.5 \times 10^2)(6 \times 10^{-7})$
$= (2.5 \times 6)(10^{2-7})$
$= 15 \times 10^{-5}$
$= 1.5 \times 10^{-4}$

2.6 Algebraic Expressions and Formulas

e.g. If $x = 4$, $y = -3$ then
$x^2 - 4y = (4)^2 - 4(-3) = 16 + 12 = 28$

e.g. Solve for L in terms of P and W
$P = 2L + 2W$
$P - 2W = 2L$
$\dfrac{P - 2W}{2} = L \qquad L = \dfrac{P - 2W}{2}$

e.g. Area of a Circle:
$A = \pi r^2$
Rearrange the formula for r in terms of A:
$r^2 = \dfrac{A}{\pi}$
$r = \sqrt{\dfrac{A}{\pi}}$

e.g. Temperature Conversion:
Given $F = \dfrac{9}{5}C + 32$
Solve for C in terms of F
$F - 32 = \dfrac{9}{5}C$
$C = \dfrac{5}{9}(F - 32)$

I. Basic Algebra

3. EQUATIONS AND INEQUALITIES

3.1 Linear Equations

$4(x + 1) = 2x + 10$ remove the ( ) first
$4x + 4 = 2x + 10$ combine variables on one side, and numbers on the other side;
$4x - 2x = 10 - 4$ change the term's sign when across the = sign
$2x = 6$
$x = 3$

e.g. Solve for x:
$2ax - 5x = 2$
$(2a - 5)x = 2$
$x = \dfrac{2}{2a - 5}$

e.g. $\dfrac{2}{3}x - 6 = \dfrac{1}{2}x + 4$
$4x - 36 = 3x + 24$ multiply both sides by LCD = 6
$4x - 3x = 24 + 36$
$x = 60$

3.2 Linear Inequalities

Solving a linear inequality is the same as solving a linear equation except when both sides of the inequality are multiplied or divided by a negative number, then the inequality sign is reversed.

e.g. $3x - 10 \geq 2$
$3x \geq 2 + 10$
$3x \geq 12$
$x \geq 4$ (solid circle • for ≥ or ≤ signs)

e.g. $3 - 2x > 9$
$-2x > 9 - 3$
$\dfrac{-2x}{-2} < \dfrac{6}{-2}$ inequality sign reversed
$x < -3$ (hollow circle o for > or < signs)

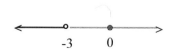

Compound Inequalities

Inequalities joined by "and" or "or"

e.g. $x > -2$ and $x < 5$
It can be written as
$-2 < x < 5$

e.g. $x < -2$ or $x > 5$

e.g. Solve $-7 \leq 2x + 5 < 11$
Rewrite it in two simple inequalities and solve them separately:

$-7 \leq 2x + 5$	$2x + 5 < 11$
$-12 \leq 2x$	$2x < 6$
$-6 \leq x$	$x < 3$

solution: $-6 \leq x < 3$

3.3 Absolute Value Equations

e.g. $|4 - x| = 3x$
remove the absolute value symbol:
$|4 - x| = \pm (4 - x)$

$4 - x = 3x$	$-(4 - x) = 3x$
$x = 1$	$x = -2$
check: $|4 - 1| = 3 \cdot 1$	check: $|4 - (-2)| \neq 3 \cdot (-2)$
(true)	(false)

e.g. $|2x + 5| - 4 = x$
remove the absolute value symbol:
$|2x + 5| = \pm (2x + 5)$

$(2x + 5) - 4 = x$	$-(2x + 5) - 4 = x$
$x = -1$	$x = -3$
(true)	(true)

I. Basic Algebra

3.4 Absolute Value Inequalities

(1) If $|x| < k$ where $k > 0$
   then $-k < x < k$

e.g. $\quad |2x + 3| < 7$
   $\quad -7 < 2x + 3 < 7$

$\quad \begin{array}{l} -7 < 2x + 3 \\ -10 < 2x \\ -5 < x \end{array} \Bigg| \begin{array}{l} 2x + 3 < 7 \\ 2x < 4 \\ x < 2 \end{array}$

solution: $-5 < x < 2$

(2) If $|x| > k$ where $k > 0$
   then $x < -k$ or $x > k$

e.g. $\quad |10 - 2x| - 2 \geq 0$
   $\quad |10 - 2x| \geq 2$
   $10 - 2x \leq -2$ or $10 - 2x \geq 2$
   $-2x \leq -12$ or $-2x \geq -8$ divided by a negative number
   $x \geq 6$ or $x \leq 4$ inequality sign reversed

3.5 Quadratic Equations

(1). Use Factoring:
e.g. $\quad x^2 - 10 = 3x$
   $\quad x^2 - 3x - 10 = 0$ set one side equal to zero
   $\quad (x + 2)(x - 5) = 0$ factor the trinomial
   $\quad x + 2 = 0$ or $x - 5 = 0$
   $\quad x = -2$ or $x = 5$
   solution set $\{-2, 5\}$

e.g. $\quad 2x^2 = 5x$
   $\quad 2x^2 - 5x = 0$
   $\quad x(2x - 5) = 0$
   $\quad x = 0$ or $2x - 5 = 0$
   $\quad \quad \quad 2x = 5$
   $\quad \quad \quad x = \dfrac{5}{2}$
   solution set $\{0, \dfrac{5}{2}\}$

e.g. $\quad x^2 + 5 = 30$
   $\quad x^2 + 5 - 30 = 0$
   $\quad x^2 - 25 = 0$
   $\quad (x + 5)(x - 5) = 0$
   $\quad x + 5 = 0$ or $x - 5 = 0$
   $\quad x = -5$ or $x = 5$
   solution set $\{-5, 5\}$

(2). Complete the Square

$$x^2 + bx + c = 0$$
$$x^2 + bx = -c$$
$$x^2 + bx + \left(\dfrac{b}{2}\right)^2 = -c + \left(\dfrac{b}{2}\right)^2$$
$$\left(x + \dfrac{b}{2}\right)^2 = -c + \left(\dfrac{b}{2}\right)^2$$

e.g. $\quad x^2 - 8x + 3 = 0$
   $\quad x^2 - 8x = -3$
   $\quad x^2 - 8x + (-4)^2 = -3 + (-4)^2$
   $\quad (x - 4)^2 = 13$
   $\quad x - 4 = \pm\sqrt{13}$
   $\quad x = 4 \pm \sqrt{13}$

(3). Quadratic Formula

$$ax^2 + bx + c = 0 \quad \text{where } a \neq 0$$
$$x = \dfrac{-b \pm \sqrt{b^2 - 4ac}}{2a}$$

e.g. $\quad 2x^2 - 4x + 1 = 0$
   $\quad a = 2, b = -4, c = 1$
   $\quad x = \dfrac{4 \pm \sqrt{(-4)^2 - 4(2)(1)}}{2(2)}$
   $\quad = \dfrac{4 \pm \sqrt{8}}{4} = \dfrac{4 \pm 2\sqrt{2}}{4}$
   $\quad = \dfrac{2 \pm \sqrt{2}}{2}$

I. Basic Algebra

(4). Roots of the Quadratic Equation

$ax^2 + bx + c = 0$ where $a \neq 0$

The sum of the roots $x_1 + x_2 = -\dfrac{b}{a}$

The product of the roots $x_1 \cdot x_2 = \dfrac{c}{a}$

e.g. Find the sum and product of the roots of the equation $3x^2 + 11x - 5 = 0$

$a = 3$, $b = 11$, $c = -5$

The sum of the roots $x_1 + x_2 = -\dfrac{b}{a} = -\dfrac{11}{3}$

The product of the roots $x_1 \cdot x_2 = \dfrac{c}{a} = -\dfrac{5}{3}$

e.g. $x_1 + x_2 = 5$ and $x_1 \cdot x_2 = 6$
Write the quadratic equation.
$5 = -\dfrac{b}{a}$, $6 = \dfrac{c}{a}$
$a = 1$, $b = -5$, $c = 6$
$x^2 - 5x + 6 = 0$

Discriminant △

$b^2 - 4ac$ is called the discriminant △
△ > 0 two unequal real roots (two x intercepts)
△ = 0 two equal real roots (one x intercept)
△ < 0 no real roots (no x intercept)

e.g. Use the discriminant to determine the properties of the roots of the equation
$3x^2 + 8x - 5 = 0$
$b^2 - 4ac = 8^2 - 4 \cdot 3 \cdot (-5) = 124$
Since 124 is greater than 0, it has two unequal real roots. And 124 is not a perfect, the roots are irrational.

e.g. Find the value of k such that the equation $4x^2 - kx + 9 = 0$ has equal roots.
Let $b^2 - 4ac = 0$
$(-k)^2 - 4 \cdot 4 \cdot 9 = 0$
$k^2 = 144$
$k = \pm 12$

3.6 Quadratic Inequalities

The solution set of a quadratic inequality is in the form of

(1) $x_1 < x < x_2$ where x_1, x_2 are the roots,
or (2) $x < x_1$ or $x > x_2$ and $x_1 < x_2$

Use the quadratic inequality to test a value from each of the three regions.

e.g. $x^2 - 7x + 10 < 0$

Solve the corresponding quadratic equation
$x^2 - 7x + 10 = 0$
$x_1 = 2$, $x_2 = 5$

test $x = 0$ (false), $x = 3$ (true), $x = 6$ (false); therefore the solution set is $2 < x < 5$.

e.g. $x^2 - 7x + 10 > 0$

Solve the corresponding quadratic equation
$x^2 - 7x + 10 = 0$
$x_1 = 2$, $x_2 = 5$

test $x = 0$ (true), $x = 3$ (false), $x = 6$ (true); therefore the solution set is $x < 2$ or $x > 5$.

I. Basic Algebra

3.7 Rational Equations

(1). Use cross-multiplication for proportions:

e.g. $\dfrac{x+2}{x-3} = \dfrac{x}{4}$

$4(x + 2) = x(x - 3)$ cross-multiply
$4x + 8 = x^2 - 3x$
$x^2 - 7x - 8 = 0$
$(x - 8)(x + 1) = 0$
$x = 8$ or $x = -1$ check: Both are solutions.

Since we use the multiplication in the process, we must check for possible extraneous solutions.

(2). Use the Least Common Denominator (LCD):

e.g. $\dfrac{2}{x^2 - x} = \dfrac{2}{x-1} + 1$ LCD $= x(x - 1)$

$2 = 2x + x(x - 1)$ multiply LCD on both sides
$2 = 2x + x^2 - x$
$x^2 + x - 2 = 0$
$(x + 2)(x - 1) = 0$
$x = -2$ or $x = 1$
check: $x = -2$ ($x = 1$ undefined)

e.g. Person A needs 20 days to complete the work and person B needs 30 days to complete the work. How many days do they need to complete the work together?

A can do $\dfrac{1}{20}$ of the work each day.

B can do $\dfrac{1}{30}$ of the work each day.

Let d be the number of the days needed.

$(\dfrac{1}{20} + \dfrac{1}{30}) \cdot d = 1$

$\dfrac{5}{60} \cdot d = 1$

$d = 12$ days

e.g.
How much 75% juice blend and how much water are needed to make 2 liters of 25% juice blend?

Let x liters be 75% juice blend.
Water will be 2 - x liters.
The key is to set the equation to pure juice amount.

$0.75x = 0.25 \cdot 2$

$x = \dfrac{2}{3}$ liters   (75% juice)

$2 - \dfrac{2}{3} = \dfrac{4}{3}$ liters   (water)

e.g.
How much 90% juice blend and 10% juice blend are needed to make 2 liters of 25% juice blend?

Let x liters be 90% juice blend.
2 - x will be 10% juice blend.
Set the equation on pure juice amount.

$0.9x + 0.1(2 - x) = 0.25 \cdot 2$
$0.9x + 0.2 - 0.1x = 0.5$
$0.8x = 0.3$

$x = \dfrac{3}{8}$ liters (90% juice)

$2 - x = \dfrac{13}{8}$ liters (10% juice)

e.g.
Find a pair of consecutive odd integers such that the sum of their reciprocals is $\dfrac{12}{35}$.

Let x be the first integer.
x + 2 will be the second integer.

$\dfrac{1}{x} + \dfrac{1}{x+2} = \dfrac{12}{35}$ LCD $= 35x(x + 2)$

$35(x + 2) + 35x = 12x(x + 2)$
$6x^2 - 23x - 35 = 0$
$(6x + 7)(x - 5) = 0$
$x = 5$ ($x = -\dfrac{7}{6}$ rejected)
$x + 2 = 7$

Solution: $\{5, 7\}$

I. Basic Algebra

3.8 Rational Inequalities

e.g. $\dfrac{x+2}{x-3} > \dfrac{x}{4}$

Solve the corresponding equation and find all undefined values of x.

Step 1. Solve the corresponding equation.
$$\dfrac{x+2}{x-3} = \dfrac{x}{4}$$
$x = 8$ or $x = -1$

Step 2. Find the undefined values of x.
$x - 3 = 0$
$x = 3$

Step 3. Divide the number line by these values.

Step 4. Test an arbitrary number from each interval into the original inequality.

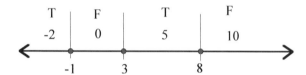

```
  T   |  F   |  T   |  F
 -2   |  0   |  5   |  10
------●------●------●--------->
     -1     3      8
```

test $x = -2$ (true), $x = 0$ (false), $x = 5$ (true)
$x = 10$ (false)
thus the solution set is $x < -1$ or $3 < x < 8$
or in Interval Notation $(-\infty, -1) \cup (3, 8)$

e.g. $\dfrac{x^2 + 2x - 11}{x^2 + x - 6} < 1$

Solve $\dfrac{x^2 + 2x - 11}{x^2 + x - 6} = 1$
$x = 5$
Find the undefined values of x.
$x^2 + x - 6 = 0$
$x = -3, \quad x = 2$

```
  T   |  F   |  T   |  F
 -5   |  0   |  3   |  10
------●------●------●--------->
     -3     2      5
```

test $x = -5$ (true), $x = 0$ (false), $x = 3$ (true)
$x = 10$ (false)
thus the solution set is $x < -3$ or $2 < x < 5$
or in Interval Notation $(-\infty, -3) \cup (2, 5)$

3.9 Radical Equations

e.g. $x = 1 + \sqrt{x+5}$
$x - 1 = \sqrt{x+5}$ isolate the radical
$(x-1)^2 = x+5$ square both sides
$x^2 - 2x + 1 = x + 5$
$x^2 - 3x - 4 = 0$
$x = 4$ or $x = -1$
must check:
plug-in the number into the original equation
$x = 1 + \sqrt{x+5}$
$x = 4:\quad 4 = 1 + \sqrt{4+5}$
$\quad\quad\quad\; 4 = 1 + \sqrt{9}$
$\quad\quad\quad\; 4 = 4$
$x = -1:\; -1 = 1 + \sqrt{-1+5}$
$\quad\quad\quad\; -1 = 1 + \sqrt{4}$
$\quad\quad\quad\; -1 = 3$ (rejected)
Therefore the solution is
$\quad\quad x = 4$

e.g. $x^{\frac{3}{2}} + 1 = 9$
$x^{\frac{3}{2}} = 8$ isolate the radical
$(x^{\frac{3}{2}})^{\frac{2}{3}} = 8^{\frac{2}{3}}$
$x^{\frac{3}{2} \cdot \frac{2}{3}} = 8^{\frac{2}{3}}$
$x^1 = 8^{\frac{2}{3}}$
$x = \sqrt[3]{8^2}$
$x = 4$
must check:
$x = 4:\quad 4^{\frac{3}{2}} + 1 = 9$
$\quad\quad\quad (\sqrt{4})^3 + 1 = 9$
$\quad\quad\quad 2^3 + 1 = 9$
$\quad\quad\quad\quad 9 = 9$
Therefore the solution is
$\quad\quad x = 4$

I. Basic Algebra

3.10 Linear System of Equations

Substitution Method:
$$2x + y = 6 \quad (1)$$
$$x = 3y + 10 \quad (2)$$
Substitute x by $3y + 10$ in Eq.(1):
$$2(3y + 10) + y = 6$$
$$6y + 20 + y = 6$$
$$6y + y = 6 - 20$$
$$7y = -14$$
$$y = -2$$
Use Eq.(2) $x = 3(-2) + 10 = -6 + 10 = 4$
Solution: $x = 4$, $y = -2$ or $(4, -2)$

Addition or Subtration Method:
$$x + y = 7 \quad (1)$$
$$3x - 2y = 1 \quad (2)$$
Multiply Eq.(1) by 2:
$$2x + 2y = 14 \quad (3)$$
Add Eq.(3) and Eq.(2)
$$2x + 2y = 14$$
$$3x - 2y = 1$$
$$\overline{}$$
$$5x = 15$$
$$x = 3$$
Substitute x by 3 in Eq. (1):
$$3 + y = 7$$
$$y = 4$$
Solution: $x = 3$, $y = 4$ or $(3, 4)$

3.11 Quadratic-Linear System

Algebraic Solution:

e.g.
$$y = x^2 - 8 \quad (1)$$
$$y + 5 = 2x \quad (2)$$

From Eq. (2) $y = 2x - 5$ (3)
Substitute y by $2x - 5$ in Eq.(1):
$$2x - 5 = x^2 - 8$$
$$x^2 - 2x - 3 = 0$$
$$(x - 3)(x + 1) = 0$$

$x - 3 = 0$	$x + 1 = 0$
$x = 3$	$x = -1$
$y = 2(3) - 5 = 1$	$y = 2(-1) - 5 = -7$

Solution: $\{(3, 1), (-1, -7)\}$

4. SEQUENCE AND SERIES

4.1 Sequence

A **sequence** is a set of numbers written in order.

A **finite sequence** has a finite number of terms.

An **infinite sequence** has an infinite number of terms.

e.g. $2, 4, 6, 8, 10, \cdots$ (infinite)

e.g. $\dfrac{1}{2}, \dfrac{1}{4}, \dfrac{1}{6}, \dfrac{1}{8}, \dfrac{1}{10}$ (finite)

The terms of a sequence are often designated as
$$a_1, a_2, a_3, a_4, \cdots, a_n, \cdots$$

e.g. $a_1 = 2$, $a_2 = 4$, $a_3 = 6$, $a_4 = 8$, $a_5 = 10$

A **sequence** is a special type of function whose domain is the set of counting numbers:

$$a_n = f(n) \quad \text{where} \quad \{n: 1, 2, 3, 4, 5, \cdots\}$$

e.g. the positive even integers:
$$a_n = 2n$$
$a_1 = 2(1) = 2$, $a_2 = 2(2) = 4$, $a_3 = 2(3) = 6, \cdots$

e.g. the positive odd integers:
$$a_n = 2n - 1$$
$a_1 = 2(1) - 1 = 1$, $a_2 = 2(2) - 1 = 3$,
$a_3 = 2(3) - 1 = 5, \cdots$

e.g. $a(n) = n^2$
$1, 4, 9, 16, 25, \cdots$

e.g. $a(n) = (-1)^n \cdot n$
$-1, 2, -3, 4, -5, 6, \cdots$

e.g. $a(n) = (-1)^{n-1} \cdot n$
$1, -2, 3, -4, 5, -6, \cdots$

I. Basic Algebra

e.g. the triangular numbers:
$$a(n) = \frac{n(n+1)}{2}$$
$$a_1 = a(1) = \frac{1(1+1)}{2} = 1$$
$$a_2 = a(2) = \frac{2(2+1)}{2} = 3$$
$$a_3 = a(3) = \frac{3(3+1)}{2} = 6$$
1, 3, 6, 10, 15, 21, • • •

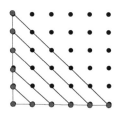

4.2 Recursive Definition

In some sequences, any term can be defined by its previous terms except the first one.

e.g. $a(n) = a(n-1) + 2$ for $n \geq 2$ and $a(1) = 5$
   $a(2) = a(2-1) + 2 = a(1) + 2 = 5 + 2 = 7$
   $a(3) = a(3-1) + 2 = a(2) + 2 = 7 + 2 = 9$
   $a(4) = a(4-1) + 2 = a(3) + 2 = 9 + 2 = 11$
   • • • • • •

e.g. $a(n) = a(n-1) + 2n$ for $n \geq 2$ and $a(1) = 5$
   $a(2) = a(2-1) + 2(2) = a(1) + 2(2) = 5 + 4 = 9$
   $a(3) = a(3-1) + 2(3) = a(2) + 2(3) = 9 + 6 = 15$
   $a(4) = a(4-1) + 2(4) = a(3) + 2(4) = 15 + 8 = 23$
   • • • • • •

e.g. $a(n+1) = 2 \cdot a(n) - n$ for $n \geq 1$ and $a(1) = 5$
   $a(2) = 2 \cdot a(1) - 1 = 2 \cdot 5 - 1 = 9$
   $a(3) = 2 \cdot a(2) - 2 = 2 \cdot 9 - 2 = 16$
   $a(4) = 2 \cdot a(3) - 3 = 2 \cdot 16 - 3 = 29$
   • • • • • •

e.g. **Fibonacci Sequence:**
   $a_1 = 1$, $a_2 = 1$
   $a_n = a_{n-2} + a_{n-1}$ for $n > 2$
   $a_3 = a_1 + a_2 = 1 + 1 = 2$
   $a_4 = a_2 + a_3 = 1 + 2 = 3$
   $a_5 = a_3 + a_4 = 2 + 3 = 5$
   • • • • • •
   1, 1, 2, 3, 5, 8, 13, 21, • • •

4.3 Arithmetic Sequences

An **arithmetic sequence** is a sequence in form of
   a, $a + d$, $a + 2d$, $a + 3d$, $a + 4d$, • • •

   $a_n = a_1 + (n-1)d$ d is the common difference
or $a_n = a_{n-1} + d$ recursive definition

e.g. { 5, 10, 15, 20, 25 }

e.g. Find the missing terms between $a_1 = 5$ and $a_6 = 15$ in an arithmetic sequence.

$$a_n = a_1 + (n-1)d$$
$$a_6 = a_1 + (6-1)d$$
$$15 = 5 + (6-1)d$$
$$d = 2$$
$$a_2 = 7, \ a_3 = 9, \ a_4 = 11, \ a_5 = 13$$

These numbers are called **arithmetic means** between a_1 and a_6.

4.4 Geometric Sequences

A **geometric sequence** is a sequence in form of
   a, ar, ar^2, ar^3, ar^4, • • •

   $a_n = a_1 \cdot r^{n-1}$ r is the common ratio
or $a_n = a_{n-1} \cdot r$ recursive definition

e.g. { 1, $\frac{1}{2}$, $\frac{1}{4}$, $\frac{1}{8}$, $\frac{1}{16}$ }

e.g. { 2, -4, 8, -16, 32, • • • }

e.g. Find the **geometric means** between $a_1 = 5$ and $a_5 = 80$.

$$a_n = a_1 \cdot r^{n-1}$$
$$a_5 = a_1 \cdot r^{5-1}$$
$$80 = 5 \cdot r^4$$
$$r^4 = 16$$
$$r = 2 \text{ or } r = -2$$
$$a_2 = 10, \ a_3 = 20, \ a_4 = 40$$
or $a_2 = -10, \ a_3 = 20, \ a_4 = -40$

I. Basic Algebra

4.5 Series

A **series** is the sum of the sequence.
Same as sequence, there are two types of series: finite series and infinite series.

A **finite series** is the sum of the finite sequence.
An **infinite series** is the sum of the infinite sequence.

e.g. $S_5 = 2 + 4 + 8 + 16 + 32$
S_5 represents the 5th **partial sum**, the sum of the first 5 terms of the sequence.

We often use the **sigma notation** $\sum$ for the sum.

e.g. $\sum_{i=1}^{5} i = 1 + 2 + 3 + 4 + 5$

here i is the **index**, **lower limit** = 1, **upper limit** = 5

e.g. $\sum_{n=1}^{5} (2n - 1) = 1 + 3 + 5 + 7 + 9$

e.g. The sum of odd numbers from 1 to 99
$$\sum_{n=1}^{50} (2n - 1)$$

e.g. The sum of even numbers from 2 to 100
$$\sum_{n=1}^{50} 2n$$

e.g. $\sum_{k=3}^{7} k^2 = 3^2 + 4^2 + 5^2 + 6^2 + 7^2$

e.g. $\sum_{n=0}^{4} (-1)^n = 1 - 1 + 1 - 1 + 1$

e.g. $\sum_{n=1}^{5} (-1)^n (2n - 1) = -1 + 3 - 5 + 7 - 9$

4.6 Arithmetic Series

An **arithmetic series** is the sum of an arithmetic sequence.

The sum of the first n terms:

$$S_n = \sum_{i=1}^{n} a_i = \frac{n}{2}(a_1 + a_n)$$

or $S_n = na_1 + \dfrac{n(n-1)d}{2}$

e.g. $1 + 2 + 3 + 4 + \cdots + 100$
$= \dfrac{100}{2}(1 + 100) = 5050$

e.g. Write the sum of the first 15 terms of the arithmetic series $1 + 4 + 7 + 10 + \cdots$ in sigma notation.

$a_1 = 1$, $d = 3$,
$a_n = a_1 + (n-1)d = 1 + 3(n-1) = 3n - 2$

$$S_{15} = \sum_{i=1}^{15} (3n - 2)$$

Find the sum of the above series.

$S_n = na_1 + \dfrac{n(n-1)d}{2}$

$= 15 \cdot 1 + \dfrac{15(15-1)3}{2}$

$= 330$

I. Basic Algebra

4.7 Geometric Series

A **geometric series** is the sum of a geometric sequence.

Since $\dfrac{(1-r^n)}{1-r} = 1 + r + r^2 + \cdots + r^{n-1}$

The sum of the first n terms:

$$S_n = \sum_{i=1}^{n} a_i = a_1 + a_1 r + a_1 r^2 + \cdots + a_1 r^{n-1}$$

$$= \dfrac{a_1(1-r^n)}{1-r}$$

e.g. $5 + \dfrac{5}{2} + \dfrac{5}{4} + \dfrac{5}{8} + \dfrac{5}{16}$

$a_1 = 5$, $r = \dfrac{1}{2}$, $n = 5$

$$S_n = \dfrac{a_1(1-r^n)}{1-r} = \dfrac{5(1-(\frac{1}{2})^5)}{(1-\frac{1}{2})} = \dfrac{5 \cdot \frac{31}{32}}{\frac{1}{2}} = \dfrac{155}{16}$$

4.8 Infinite Series

1. An infinite arithmetic series has no limit.
2. An infinite geometric series has no limit when $|r| \geq 1$.
3. An infinite geometric series has a limit when $|r| < 1$.

$$S_n = \sum_{i=1}^{\infty} a_i = \dfrac{a_1}{1-r}$$

e.g. Find the limit of $0.252525\cdots$

$0.252525\cdots = 0.25 + 0.0025 + 0.000025 + \cdots$

$a_1 = 0.25$, $r = \dfrac{1}{100} = 0.01$

$$S_n = \dfrac{a_1}{1-r} = \dfrac{0.25}{1-0.01} = \dfrac{0.25}{0.99} = \dfrac{25}{99}$$

e.g. Find the limit of the infinite geometric series:

$1 + \dfrac{1}{2} + \dfrac{1}{4} + \dfrac{1}{8} + \dfrac{1}{16} + \cdots$

$a_1 = 1$, $r = \dfrac{1}{2} = 0.5$

$$S_n = \dfrac{a_1}{1-r} = \dfrac{1}{1-0.5} = \dfrac{1}{0.5} = 2$$

e.g. Find the limit of the infinite geometric series:

$1 - \dfrac{1}{2} + \dfrac{1}{4} - \dfrac{1}{8} + \dfrac{1}{16} + \cdots$

$a_1 = 1$, $r = -\dfrac{1}{2} = -0.5$

$$S_n = \dfrac{a_1}{1-r} = \dfrac{1}{1+0.5} = \dfrac{1}{1.5} = \dfrac{2}{3}$$

4.9 The Number e

$$\sum_{n=0}^{\infty} \dfrac{1}{n!} = 1 + \dfrac{1}{1!} + \dfrac{1}{2!} + \dfrac{1}{3!} + \cdots = e$$

$e = 2.718281828\cdots$

II. Basic Geometry

5. LOGIC

5.1 Negation: not, Symbol ~
e.g. Statement p: I am a student.   T
      Negation ~ p: I am not a student.   F

Truth Values:

p	~p
T	F
F	T

5.2 Conjunction: and, Symbol ∧
The conjunction p ∧ q is true only when both parts are true.

Truth Values

p	q	p ∧ q
T	**T**	**T**
T	F	F
F	T	F
F	F	F

5.3 Disjunction: or, Symbol v
The disjunction p v q is false only when both parts are false.

Truth Values

p	q	p v q
T	T	T
T	F	T
F	T	T
F	**F**	**F**

e.g. p: 10 is divisible by 2.   T
     q: 10 is divisible by 3.   F
     p ∧ q: 10 is divisible by 2 and 10 is divisible by 3.   F
     p v q: 10 is divisible by 2 or 10 is divisible by 3.   T
     p ∧ ~q: 10 is divisible by 2 and 10 is not divisible by 3.   T

5.4 Conditional Statements
(1) **Original**   p ----> q   (If p, then q.)
e.g. If it is snowing, then the school is closed.
(2) **Inverse** ~p ---> ~q (If not p, then not q.)
e.g. If it is not snowing, then the school is not closed.
(3) **Converse** q ---> p   (If q, then p.)
e.g. If the school is closed, then it is snowing.
(4) **Contrapositive** ~q ---> ~p (If not q, then not p.)
 e.g. If the school is not closed, then it is not snowing.

Statements (1) and (4) are logically equivalent.
Statements (2) and (3) are logically equivalent.

5.5 Biconditional Statements
p <----> q (p and q have the same truth value.)
e.g. All the definitions are biconditional statements.
e.g. Two lines are perpendicular if and only if they form right angles.

6. POSTULATES

6.1 Postulates of Equality
(1)   a = a   **Reflexive Property**

e.g.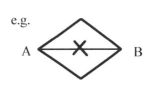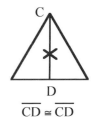

$\overline{AB} \cong \overline{AB}$ $\overline{CD} \cong \overline{CD}$

(2)   If a = b, then b = a   **Symmetric Property**
(3)   If a = b and b = c   **Transitive Property**
      then a = c
(4)   If a = f(b) and b = c   **Substitution Property**
      then a = f(c)
e.g. If a = 2b and b = c
      then a = 2c
(5)   If a = b and c = d   **Addition Property**
      then a + c = b + d
e.g. If a = b then a + c = b + c
(6)   If a = b and c = d   **Multiplication Property**
      then ac = bd

e.g. If a = b then $\dfrac{a}{2} = \dfrac{b}{2}$ Halves of equal quantities are equal (**Division Property**)

6.2 Postulates of Inequality
(1)   If a > b and b > c   **Transitive Property**
      then a > c

(2)   If a > b and b = c   **Substitution Property**
      then a > c

(3)   If a > b and c > d   **Addition Property**
      then a + c > b + d
e.g. If a > b then a + c > b + c

(4)   If a > b and c > 0   **Multiplication Property**
      then ac > bc
e.g. If a > b then 2a > 2b

6.3 Partition Postulate
A whole is equal to the sum of all its parts.
e.g AD = AB + BC + CD
A whole is greater than any of its parts.
e.g. AD > AB , AD > BD , AD > BC

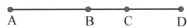

A   B   C   D

15.

II. Basic Geometry

7. DEFINITIONS

7.1 Angles

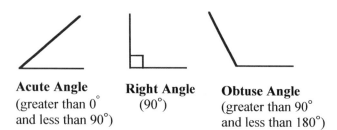

Acute Angle
(greater than 0°
and less than 90°)

Right Angle
(90°)

Obtuse Angle
(greater than 90°
and less than 180°)

If ∠A and ∠B are **complementary**, then
 m∠A + m∠B = 90   vice versa.
If ∠A and ∠B are **supplementary**, then
 m∠A + m∠B = 180   vice versa.
e.g. A linear pair of angles are supplementary.

7.2 A **midpoint** divides a line segment into two congruent segments.

7.3 A **bisector of a segment** divides the segment into two congruent segments.

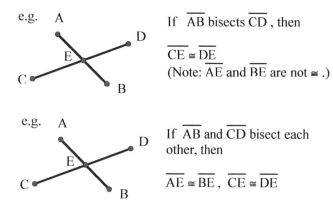

If $\overline{AB}$ bisects $\overline{CD}$, then

$\overline{CE} \cong \overline{DE}$
(Note: $\overline{AE}$ and $\overline{BE}$ are not ≅.)

If $\overline{AB}$ and $\overline{CD}$ bisect each other, then

$\overline{AE} \cong \overline{BE}$, $\overline{CE} \cong \overline{DE}$

7.4 A **bisector of an angle** divides the angle into two congruent angles.

7.5 Perpendicular lines intersect to form right angles.

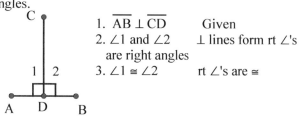

1. $\overline{AB} \perp \overline{CD}$ Given
2. ∠1 and ∠2   ⊥ lines form rt ∠'s
 are right angles
3. ∠1 ≅ ∠2   rt ∠'s are ≅

7.6 Parallel lines are in the same plane and do not intersect.

8. THEOREMS

8.1 Congruent Angles
Vertical angles are congruent.

e.g.
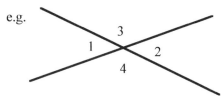

∠1 ≅ ∠2 and ∠3 ≅ ∠4   (vertical angles are ≅.)
m∠1 + m∠4 = 180   and   m∠2 + m∠4 = 180
(A linear pair of angles are supplementary.)

All right angles are congruent.

If two angles are congruent, then their complements are congruent.

If two angles are congruent, then their supplements are congruent.

8.2 Properties of Perpendicular Lines
(1) Perpendicular lines form right angles.
(2) If two lines intersect to form congruent adjacent angles, then they are perpendicular.
(3) If a point is on the perpendicular bisector of a line segment, then it is equidistant from the endpoints of the line segment, and vice versa.
(4) If two points are each equidistant from the endpoints of a line segment, these points determine the perpendicular bisector of the segment.

8.3 Properties of Parallel Lines
Parallel lines are equidistant everywhere.

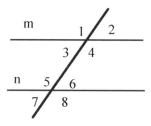

If line m ‖ line n, then alternate interior angles are ≅.
 ∠3 ≅ ∠6 and ∠4 ≅ ∠5
If line m ‖ line n, then corresponding angles are ≅.
 ∠1 ≅ ∠5, ∠2 ≅ ∠6, ∠3 ≅ ∠7, ∠4 ≅ ∠8
If line m ‖ line n, then interior angles on the same side of the transversal are supplementary.
 m∠3 + m∠5 = 180
 m∠4 + m∠6 = 180

II. Basic Geometry

9. TRIANGLES AND PROOFS

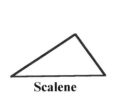

Scalene   **Isosceles**   **Equilateral**
(no congruent sides) (2 congruent sides) (3 congruent sides)

9.1 The sum of the three interior angles of a $\triangle$ is 180°. The exterior angle is equal to the sum of 2 nonadjacent interior angles.

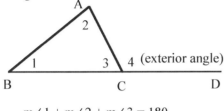

$m\angle 1 + m\angle 2 + m\angle 3 = 180$

$m\angle 1 + m\angle 2 = m\angle 4$

9.2 Triangle Inequalities

(1) In any triangle the greater side is opposite the greater angle, and vice versa.

(2) Any side is greater than the difference of the other 2 sides and less than the sum of them.
$$|s_1 - s_2| < s_3 < |s_1 + s_2|$$
e.g. If the two sides of a triangle are 3 and 5, then the 3rd side s_3 is $|3 - 5| < s_3 < 3 + 5$, which is $2 < s_3 < 8$

(3) Any exterior angle is greater than either nonadjacent interior angle.

9.3 Median, Altitude, and Angle Bisector

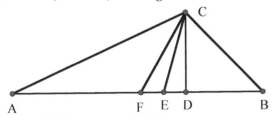

Given: $\overline{CD}$ is an altitude, $\overline{CE}$ is an angle bisector, and $\overline{CF}$ is a median.

$\overline{CD} \perp \overline{AB}$	Def. of altitude
$\angle ACE \cong \angle BCE$	Def. of angle bisector
F is the midpoint of $\overline{AB}$	Def. of median
$\overline{AF} \cong \overline{BF}$	Def of midpoint

Concurrence (Intersect in One Point)

(1) Centroid:
The medians of a triangle are concurrent.
(It is the center of gravity.)
The centroid divides each median in the ratio 2 to 1.

e.g.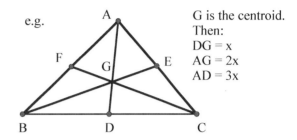

G is the centroid.
Then:
$DG = x$
$AG = 2x$
$AD = 3x$

(2) Orthocenter:
The altitudes of a triangle are concurrent.
The orthocenter of an obtuse $\triangle$ is outside of the triangle.

(3) Incenter:
The angle bisectors of a triangle are concurrent.
(It is the center of the inscribed circle
--- equidistant from each side.)

(4) Circumcenter:
The perpendicular bisectors of the sides of a triangle are concurrent.
(It is the center of the circumscribed circle
--- equidistant from each vertex.)
The circumcenter of an obtuse $\triangle$ is outside of the triangle.

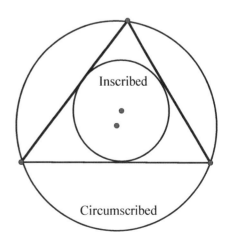

Inscribed

Circumscribed

II. Basic Geometry

9.4 Right Triangle

Pythagorean Theorem

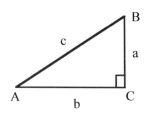

∠C is a right angle
a and b are legs
c is the hypotenuse

$$a^2 + b^2 = c^2$$

Pythagorean Triples:
  3, 4, 5;   6, 8, 10;   9, 12, 15 etc.
  5, 12, 13;   10, 24, 26 etc.

e.g.
The ratio of two legs are 3:4 and the hypotenuse is 15.
Find the lengths of the two legs:
$$(3n)^2 + (4n)^2 = 15^2$$
$$9n^2 + 16n^2 = 15^2$$
$$25n^2 = 225$$
$$n^2 = 9$$
$$n = 3 \quad (n = -3 \text{ rejected})$$
$3n = 3 \cdot 3 = 9$ and $4n = 4 \cdot 3 = 12$
The lengths of the two legs are 9 and 12.

Theorem: In a right triangle, the median to the hypotenuse is equal to half of the hypotenuse.

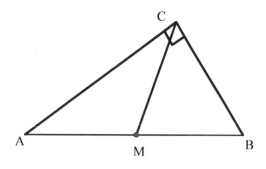

M is the midpoint.

CM = AM = BM

Special Right Triangles

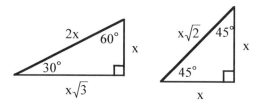

9.5 Isosceles Triangle

Definition: A triangle that has two congruent sides.

(1) The base angles of an isosceles triangle are congruent.
(2) If two angles of a triangle are congruent, then their opposite sides are congruent.
(3) To the base of an isosceles triangle, the median, altitude, angle bisector, and perpendicular bisector coincide.

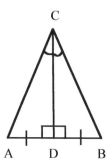

△ABC is isosceles with $\overline{AC} \cong \overline{BC}$, $\overline{CD}$ is the median, altitude, angle bisector, and perpendicular bisector.

9.6 Equilateral Triangle

Definition: A triangle that has three congruent sides.

An equilateral triangle is equiangular, each angle is 60°.

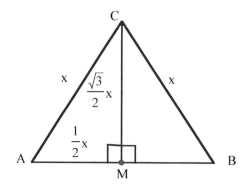

II. Basic Geometry

9.7 Congruent Triangles and Proofs

If two triangles are congruent, then their corresponding angles and corresponding sides are congruent.

CPCTC: Corresponding Parts of Congruent Triangles are Congruent.

Proving Triangles Congruent:

S.S.S. ≅ S.S.S.   Side-Side-Side Congruent
S.A.S. ≅ S.A.S.   Side-Angle-Side Congruent
A.S.A. ≅ A.S.A.   Angle-Side-Angle Congruent
A.A.S. ≅ A.A.S.   Angle-Angle-Side Congruent

Note: S.S.A. Side-Side-Angle Congruent can not be used to prove two triangles congruent.

For right triangles only, we can also use
H.L. ≅ H.L.   Hypotenuse-Leg Congruent

e.g. $\overline{AB}$ and $\overline{CD}$ bisect each other at point E.
Prove: △ACE ≅ △BDE

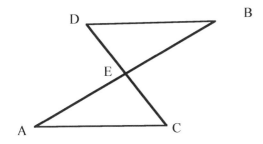

1. $\overline{AB}$ and $\overline{CD}$ bisect each other at point E   1. Given
2. $\overline{AE} \cong \overline{BE}$ and $\overline{CE} \cong \overline{DE}$ 2. A bisector divides a segment into two congruent segments
3. ∠AEC ≅ ∠BED   3. Vertical angles are congruent
4. △ACE ≅ △BDE   4. S.A.S. ≅ S.A.S.

e.g. $\overline{AC}$ bisects ∠BAD and ∠BCD.
Prove: $\overline{AB} \cong \overline{AD}$

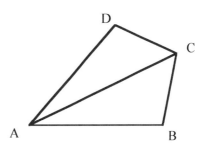

1. $\overline{AC}$ bisects ∠BAD and ∠BCD   1. Given
2. ∠BAC ≅ ∠DAC
   ∠BCA ≅ ∠DCA   2. Def. of angle bisector
3. $\overline{AC} \cong \overline{AC}$ 3. Reflexive property
4. △ABC ≅ △ADC   4. A.S.A. ≅ A.S.A.
5. $\overline{AB} \cong \overline{AD}$ 5. CPCTC

e.g. $\overline{CA} \perp \overline{AB}$, $\overline{DB} \perp \overline{AB}$, $\overline{AD} \cong \overline{BC}$
Prove: △ABC ≅ △BAD

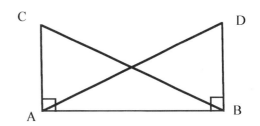

1. $\overline{CA} \perp \overline{AB}$, $\overline{DB} \perp \overline{AB}$ 1. Given
2. ∠CAB and ∠DBA are right angles   2. ⊥ lines form right angles
3. △ABC and △BAD are right △s   3. Def. of the right △
4. $\overline{AD} \cong \overline{BC}$ 4. Given
5. $\overline{AB} \cong \overline{AB}$ 5. Reflexive property
6. △ABC ≅ △BAD   6. H.L. ≅ H.L.

19.

9.8 Similar Triangles, Ratios and Proportions

(1) Ratios and Proportions

If two ratios are equal, they are in proportion.

$$\frac{a}{b} = \frac{c}{d} \quad \text{or} \quad a \cdot d = b \cdot c$$

In a proportion, the product of the means is equal to the product of the extremes.

(2) Similar Triangles

If two triangles are similar, then their corresponding angles are congruent and their corresponding sides are in proportion.

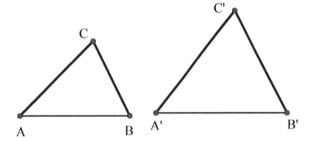

If $\triangle ABC \sim \triangle A'B'C'$, then
$\angle A \cong \angle A'$, $\angle B \cong \angle B'$, $\angle C \cong \angle C'$
$$\frac{AB}{A'B'} = \frac{BC}{B'C'} = \frac{CA}{C'A'}$$

Proving Triangles Similar:

A.A. ~ Two pairs of corresponding angles are congruent (most often used for proof)
S.A.S. ~ One pair of corresponding angles congruent and the two pairs of corresponding adjacent sides in proportion
S.S.S. ~ Three pairs of corresponding sides in proportion

Symmetric Property:

If $\triangle ABC \sim \triangle A'B'C'$ then
$\triangle A'B'C' \sim \triangle ABC$

Transitive Property:

If $\triangle I \sim \triangle II$ and $\triangle II \sim \triangle III$, then
$\triangle I \sim \triangle III$

In two similar triangles, the ratio of the perimeters is equal to the ratio of the sides.

In two similar triangles, the ratio of the areas is equal to the square of the ratio of the sides.

e.g If $\triangle ABC \sim \triangle A'B'C'$ and $\dfrac{AB}{A'B'} = \dfrac{2}{1}$

then $\dfrac{P}{P'} = \dfrac{2}{1}$ and $\dfrac{\text{Area}}{\text{Area}'} = \dfrac{4}{1}$

Theorem

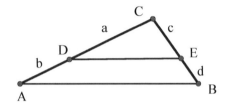

If $\overline{DE} \parallel \overline{AB}$, then $\dfrac{a}{b} = \dfrac{c}{d}$,
$$\dfrac{CD}{AC} = \dfrac{CE}{BC}$$

Angle Bisector Theorem

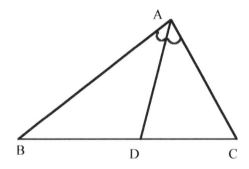

If $\angle BAD \cong \angle CAD$, then
$$\dfrac{AB}{AC} = \dfrac{BD}{DC}$$

II. Basic Geometry

Midsegment Theorem

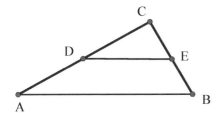

If D and E are midpoints of $\overline{AC}$ and $\overline{BC}$, then the midsegment $\overline{DE}$ is parallel to $\overline{AB}$ and is half of $\overline{AB}$.

e.g.

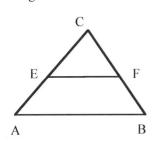 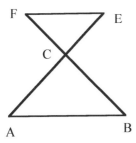

If $\overline{EF} \parallel \overline{AB}$,
then $\triangle EFC \sim \triangle ABC$ AA~
then $\dfrac{EC}{AC} = \dfrac{FC}{BC} = \dfrac{EF}{AB}$ Corresponding sides in proportion

e.g.

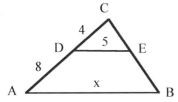

If $\overline{AB} \parallel \overline{DE}$
then $\dfrac{5}{x} = \dfrac{4}{4+8}$, $x = 15$

(Note: $\dfrac{5}{x} \neq \dfrac{4}{8}$)

(3) Proportions in the Right Triangle

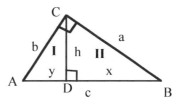

Right Triangle Altitude Theorem

(a). $\triangle I \sim \triangle II \sim \triangle ABC$

(b). The altitude to the hypotenuse is the geometric mean of the two segments of the hypotenuse.
$$h^2 = xy$$

(c). Each leg is the geometric mean of its projection on the hypotenuse and the whole hypotenuse.
$$a^2 = x(x+y) = xc$$
$$b^2 = y(x+y) = yc$$

10. INDIRECT PROOF

(1) Assume that the opposite of the conclusion is true.
(2) Show that the assumption contradicts a known fact.
(3) Since the assumption is false, the conclusion is true.

e.g.

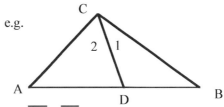

Given: $\overline{AC} \cong \overline{BC}$ and $\angle 1 \neq \angle 2$
Prove: $\overline{CD}$ is not a median

(1) $\overline{AC} \cong \overline{BC}$, $\angle 1 \neq \angle 2$ Given
(2) $\overline{CD}$ is a median   Assumed
(3) D is the midpoint   Def. of median
(4) $\overline{AD} \cong \overline{BD}$ Def. of midpoint
(5) $\overline{CD} \cong \overline{CD}$ Reflexive Property
(6) $\triangle ACD \cong \triangle BCD$ S.S.S $\cong$
(7) $\angle 1 \cong \angle 2$ CPCTC
(8) $\overline{CD}$ is not a median   Contradiction in (7) and (1)

II. Basic Geometry

11. POLYGONS

11.1 Quadrilateral

A 4-sided polygon.

11.2 To Prove a Parallelogram

2 pairs of opposite sides are parallel;
2 pairs of opposite sides are congruent;
2 pairs of opposite angles are congruent;
1 pair of opposite sides are parallel and congruent;
Diagonals bisect each other.

Rhombus: All the properties of a parallelogram;
4 sides are congruent;
Diagonals are perpendicular;
Diagonals bisect the interior angles.

Rectangle: All the properties of a parallelogram;
4 right angles;
Diagonals are congruent.

Square: All the properties of a rhombus and a rectangle.

11.3 Trapezoid

One and only one pair of opposite sides are parallel.
The median of a trapezoid is parallel to the bases.
The length of the median is equal to one-half the sum of the lengths of the bases.

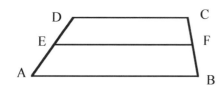

$\overline{EF}$ is the median: $\overline{EF} \parallel \overline{AB}$, $\overline{EF} \parallel \overline{CD}$

$$EF = \frac{AB + CD}{2}$$

Isosceles Trapezoid: the nonparallel sides are congruent.
Base angles are congruent.
Diagonals are congruent.

In general, parallel lines cut transversals into proportional segments.

If $l_1 \parallel l_2 \parallel l_3$, then

$$\frac{x}{y} = \frac{x'}{y'}$$

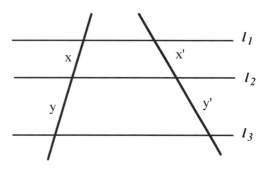

11.4 Polygons

Square (4 sides), Pentagon (5 sides), Hexagon (6 sides), Octagon (8 sides).

Sum of the exterior angles = 360°

Sum of the interior angles = n•180° - 360°
= (n - 2)•180°

A **regular polygon** has congruent sides and congruent angles.

Exterior angle of a regular polygon = $\frac{360°}{n}$

Interior angle of a regular polygon = $180° - \frac{360°}{n}$

Interior angle of a square = 90°
Interior angle of a regular pentagon = 108°
Interior angle of a regular hexagon = 120°
Interior angle of a regular octagon = 135°

II. Basic Geometry

12. CIRCLE

12.1 Angles of a Circle

The degree measure of a circle is 360°.
The degree measure of a semicircle is 180°.

Interior Cases:

The degree measure of an arc is equal to the measure of the central angle that intercepts the arc.

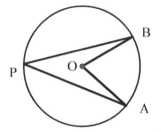

Central angle m∠AOB = m$\widehat{AB}$

Inscribed Angle Theorem:

Inscribed angle m∠APB = $\frac{1}{2}$m$\widehat{AB}$

Thales' Theorem:

Inscribed angle of a semicircle is a right angle, vice versa.

A line segment that connects any two distinct points on the circle is called a **chord**.

A line that intersects a circle at exactly one point is called a **tangent line**.

A line that intersects a circle at exactly two points is called a **secant line**.

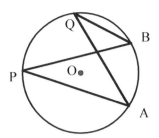

Inscribed angles that intercept the same arc are equal in measure.

$$m\angle P = m\angle Q$$

(Both of them intercept arc $\widehat{AB}$)

Two arcs are similar if they have the same angle measure.

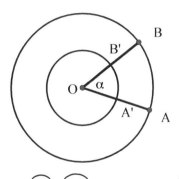

$\widehat{AB} \sim \widehat{A'B'}$
Both of them have the same angle measure α°.

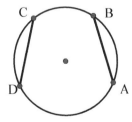

Congruent arcs have congruent chords, and vice versa.
If $\widehat{AB} \cong \widehat{CD}$, then $\overline{AB} \cong \overline{CD}$, and vice versa.

II. Basic Geometry

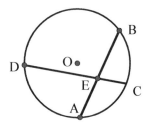

Chord - chord angle: $m\angle BEC = \frac{1}{2}(m\widehat{BC} + m\widehat{AD})$;

$m\angle AEC = \frac{1}{2}(m\widehat{AC} + m\widehat{BD})$

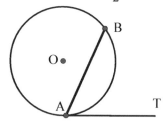

Chord - tangent angle: $m\angle BAT = \frac{1}{2}m\widehat{AB}$

Exterior Cases:

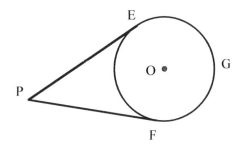

Tangent - tangent angle:
$m\angle EPF = \frac{1}{2}(m\widehat{EGF} - m\widehat{EF})$

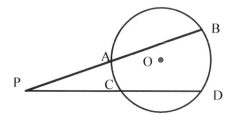

Secant - secant angle:
$m\angle BPD = \frac{1}{2}(m\widehat{BD} - m\widehat{AC})$

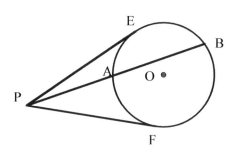

Tangent - secant angle:
$m\angle EPB = \frac{1}{2}(m\widehat{EB} - m\widehat{EA})$

$m\angle FPB = \frac{1}{2}(m\widehat{FB} - m\widehat{FA})$

e.g.

Given: $\overline{PE}$ is a tangent segment.
$m\widehat{EA} : m\widehat{AB} : m\widehat{BE} = 2 : 3 : 4$

Find: $m\angle P$, $m\angle PEA$, $m\angle PAE$

$m\widehat{EA} : m\widehat{AB} : m\widehat{BE} = 2x : 3x : 4x$
$2x + 3x + 4x = 360 \qquad x = 40$

$m\widehat{EA} = 80$, $m\widehat{AB} = 120$, $m\widehat{BE} = 160$

$m\angle P = \frac{1}{2}(m\widehat{BE} - m\widehat{EA}) = 40$

$m\angle PEA = \frac{1}{2}m\widehat{EA} = 40$

$m\angle PAE = 180 - 40 - 40 = 100$

II. Basic Geometry

12.2 Segments of a Circle

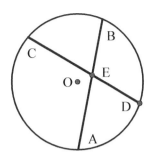

AE • EB = CE • ED

or $\dfrac{AE}{CE} = \dfrac{ED}{EB}$

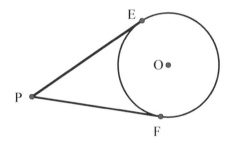

$\overline{PE}$, $\overline{PF}$ are tangent segments.

$\overline{PE} \cong \overline{PF}$

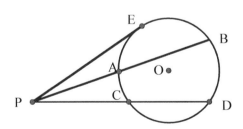

$\overline{PE}$ is a tangent segment.

$PE^2 = PA \cdot (PA + AB) = PC \cdot (PC + CD)$
or
$PE^2 = PA \cdot PB = PC \cdot PD$

12.3 Theorems

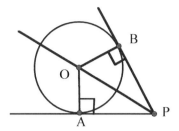

A tangent to a circle is perpendicular to the radius at its point of intersection.

$\overrightarrow{PO}$ is the angle bisector of the tangent-tangent angle ∠APB.

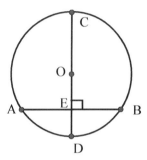

A diameter perpendicular to a chord bisects the chord and its arcs, and vice versa.

If $\overline{AB} \perp \overline{CD}$, then $\overline{AE} \cong \overline{BE}$ and $\overset{\frown}{AD} \cong \overset{\frown}{BD}$, $\overset{\frown}{AC} \cong \overset{\frown}{BC}$.

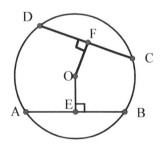

If two chords of a circle are congruent, then they are equidistant from the center of the circle, and vice versa.

If $\overline{AB} \cong \overline{CD}$, then $OE = OF$;   or
If $OE = OF$, then $\overline{AB} \cong \overline{CD}$
If $OE < OF$, then $AB > CD$

25.

II. Basic Geometry

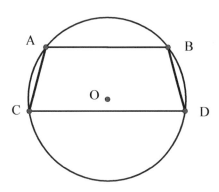

In a circle, parallel chords intercept congruent arcs between them.
If $\overline{AB} \parallel \overline{CD}$, then $\overset{\frown}{AC} \cong \overset{\frown}{BD}$

12.4 Common Tangents of Two Circles

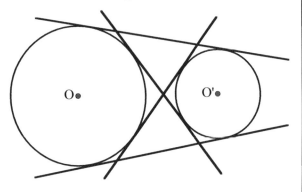

12.5 Arc Length and Area of Sector

We can use proportions to find the arc length and the area of sector in a circle.

Arc Length: $\quad s = \dfrac{x°}{360°} \cdot 2\pi r$

Area of Sector: $\quad A = \dfrac{x°}{360°} \cdot \pi r^2$

where $x°$ is the angle measure of the arc.

12.6 Inscribed Circle

A circle is inscribed in a polygon if each side of the polygon is tangent to the circle.

12.7 Inscribed Polygon and Circumscribed Circle

A polygon is inscribed in a circle if all the vertices of the polygon are on the circle. The circle is called the **circumscribed circle**.

An inscribed quadrilateral is also called a **cyclic quadrilateral**.

A quadrilateral is cyclic if and only if its opposite angles are supplementary.

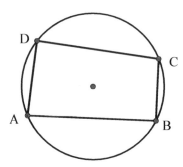

$m\angle A + m\angle C = m\angle B + m\angle D = 180°$

Special Case:

A rectangle is cyclic.
The intersecting point of the diagonals of a rectangle is the center of the circle it inscribed.

Ptolemy's Theorem:

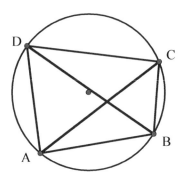

$AC \cdot BD = AB \cdot CD + BC \cdot AD$

26.

II. Basic Geometry

13. CONSTRUCTIONS AND LOCI

13.1 Three Types of Constructions

(1) Construct a line segment or an angle congruent to a given line segment or angle:

Conguent segments have the same length.

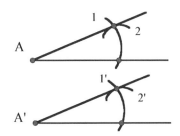

Conguent angles have the same measure.

(2) Bisect a line segment or an angle:

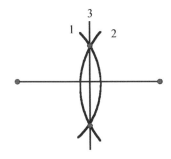

Two points equidistant from the endpoints of a line segments determine the perpendicular bisector.

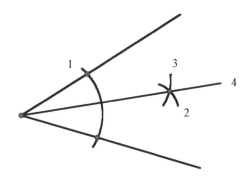

Construct two congruent △'s. (SSS)
Corresponding angles are congruent. (CPCTC)

(3) Through a point draw a line ⊥ or ∥ to a given line:

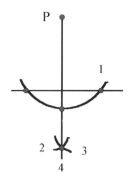

Two points equidistant from the endpoints of a line segments determine the perpendicular bisector.

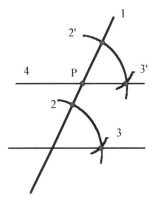

Two lines are parallel if corresponding angles are ≅.

II. Basic Geometry

13.2 Five Fundamental Loci

The **locus of points** is the set of all points satisfying a given condition or conditions.

(1) The locus of points equidistant from a given point:

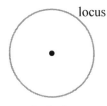

A circle

(2) The locus of points equidistant from two given points:

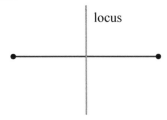

A perpendicular bisector

(3) The locus of points equidistant from two sides of a given angle:

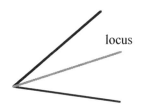

An angle bisector

(4) The locus of points equidistant from a given line:

A pair of parallel lines

(5) The locus of points equidistant from two given parallel lines:

A parallel line midway between the given lines

13.3 Compound Loci

Find the points of intersection of different loci.

e.g. How many points are 2 units from a given line and 3 units from a given point on the given line?

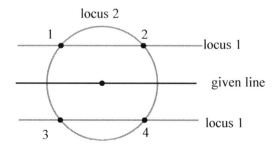

There are four points of intersection.

e.g. Find the points that are 3 units from the origin and equidistant from the x and y axes.

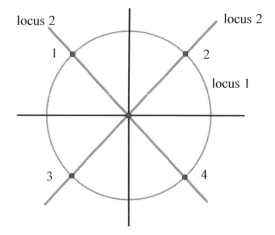

There are four points of intersection.

28.

II. Basic Geometry

14. TRANSFORMATIONS

A transformation changes the position of the points in a plane. A transformation is a mapping, a one-to-one function.

14.1 Transformation Rules

(1). Line Reflection:
A line reflection gives out the mirror image of the original object.

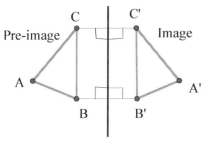

m Line of Reflection

The line of reflection *m* is the perpendicular bisector of the segments of $\overline{AA'}$, $\overline{BB'}$ and $\overline{CC'}$.

$P(x, y) \quad \underline{\text{r x-axis}} \quad P'(x, -y)$
$P(x, y) \quad \underline{\text{r y-axis}} \quad P'(-x, y)$
$P(x, y) \quad \underline{\text{r y = x}} \quad P'(y, x)$
$P(x, y) \quad \underline{\text{r y = - x}} \quad P'(-y, -x)$

(2). Point Reflection about the Origin:
A point reflection is the 180° rotation about the origin.

$P(x, y) \quad \underline{\text{r o}} \quad P'(-x, -y)$

(3). Translation:
A translation shifts the points left, right, up and down. In a translation all the points in the plane move the same distance and in the same direction.

$P(x, y) \quad \underline{\text{T a, b}} \quad P'(x + a, y + b)$

(4). Rotation about the Origin:
In a rotation all points in the plane rotate the same degrees of the angle measure about the same point - the **center of rotation**.

A counterclockwise rotation is positive.
A clockwise rotation is negative.

$P(x, y) \quad \underline{\text{R 90°}} \quad P'(-y, x)$
$P(x, y) \quad \underline{\text{R 180°}} \quad P'(-x, -y)$
$P(x, y) \quad \underline{\text{R -90°}} \quad P'(y, -x)$

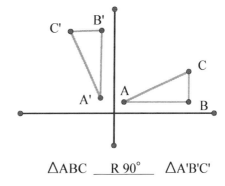

$\triangle ABC \quad \underline{\text{R 90°}} \quad \triangle A'B'C'$

(5). Dilation about the Origin:
In a dilation the image is similar to the original figure. A dilation maps angles to corresponding angles of equal measure and sides to corresponding sides in proportion.

$P(x, y) \quad \underline{\text{D o, k}} \quad P'(kx, ky)$

Only dilation enlarges or reduces the size of the image, which is similar to the original. The image of other transformations is congruent to the original.

Scale Drawing:

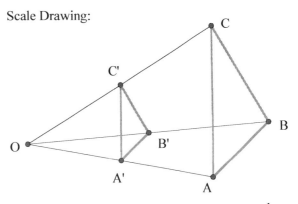

Construct the image with a scale factor $k = \frac{1}{2}$.
A', B', C' are midpoints of $\overline{OA}$, $\overline{OB}$, $\overline{OC}$.

II. Basic Geometry

e.g. Divide a segment $\overline{AB}$ into 3 equal parts.

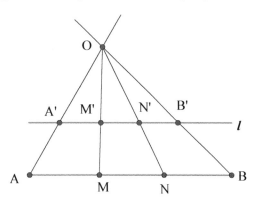

1. Construct line $l \parallel \overline{AB}$.
2. Construct $\overline{A'M'} \cong \overline{M'N'} \cong \overline{N'B}$ on line l.
3. Construct $\overrightarrow{AA'}$ and $\overrightarrow{BB'}$ intersecting at point O.
4. Construct $\overrightarrow{OM'}$ and $\overrightarrow{ON'}$ intersecting $\overline{AB}$ at point M and point N.
5. $\overline{AM} \cong \overline{MN} \cong \overline{NB}$

It is not practical to memorize these transformation rules. One should be able to derive these rules through the drawings.

e.g. Use point P(5, 1) in the drawing to derive the transformation rules.

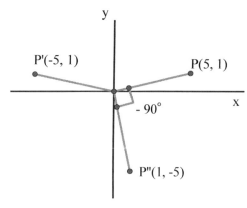

P(5, 1)   r y-axis   P'(-5, 1)
Derive: P(x, y)   r y-axis   P'(-x, y)

P(5, 1)   R -90°   P''(1, -5)
Derive: P(x, y)   R -90°   P''(y, -x)

e.g. The difference between translation and rotation.

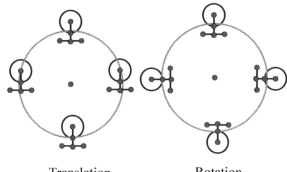

Translation   Rotation

14.2 Composition of Transformations

e.g.   r x-axis ∘ r y-axis (x, y)
     = r x-axis (-x, y) = (-x, -y)
     We can see r x-axis ∘ r y-axis = r o

e.g. $D_4 \circ T_{3, 0}(x, y) = D_4(x + 3, y) = (4x + 12, 4y)$

or (x, y) $\xrightarrow{T_{3,0}}$ (x + 3, y) $\xrightarrow{D_4}$ (4x + 12, 4y)

Glide Reflection:

Glide Reflection is a special composition of reflections and translations: $T_{a, 0} \circ r_{\text{x-axis}}$.

e.g. Find the image of P(1, 2) under the glide reflection of $T_{2,0} \circ r_{\text{x-axis}}$

P(1, 2) $\xrightarrow{\text{r x-axis}}$ P'(1, -2) $\xrightarrow{T_{2,0}}$ P''(3, -2)

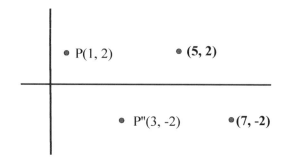

II. Basic Geometry

14.3 Symmetry

(1). Line Symmetry:

e.g.

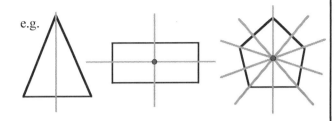

Isosceles triangle has 1 axis of symmetry;
Rectangle has 2 axes of symmetry;
Regular pentagon has 5 axes of symmetry.

(2). Point Symmetry:

e.g.

(3). Rotational Symmetry:

e.g.

Equilateral triangle has 120° rotational symmetry;
Square has 90° rotational symmetry;
Rectangle has 180° rotational symmetry;
Regular pentagon has 72° rotational symmetry;
Regular hexagon has 60° rotational symmetry.

Identity Symmetry:
Any figure has 360° rotational symmetry.

14.4 Rigid Motion and Orientation

Rigid Motion: A transformation that preserves the distance between points (both the length of the segment and the measure of the angle are unchanged).

A rigid motion is also called an **isometry**.

An object under a rigid motion is congruent to the original object.

Reflection, translation, and rotation are rigid motions. Only dilation is not a rigid motion.

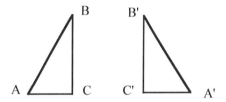

From A ---> B ---> C is **clockwise orientation**.
From A' ---> B' ---> C' is **counterclockwise orientation**.

Direct Isometry: An isometry preserves orientation;
Opposite Isometry: An isometry changes orientation.

e.g. Line Reflection: Opposite Isometry
     Point Reflection: Direct Isometry
     Rotation: Direct Isometry
     Translation: Direct Isometry
     Dilation: Changes size --- Not an isometry,
           but preserves orientation

The composition of a direct isometry and an opposite isometry is an opposite isometry.

The composition of two opposite isometries is a direct isometry.

e.g. P(x, y) __r y-axis__ P'(-x, y) __r x-axis__ P''(-x, -y)
    same as P(x, y) __R 180°__ P''(-x, -y)

Congruence is a composition of rigid motions.
Similarity is a composition of dilations and/or rigid motions.

II. Basic Geometry

15. SOLID GEOMETRY

15.1 To Determine a Plane

(1) Three noncollinear points determine a plane.

(2) Two intersecting lines determine a plane.

(3) Two parallel lines determine a plane.

Skew lines are neither parallel nor intersecting, they are not in a same plane --- not coplanar.

e.g. In a triangular right prism

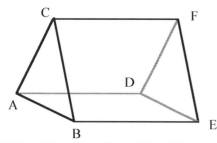

$\overline{AB} \parallel \overline{DE}$, $\overline{BC} \parallel \overline{EF}$, $\overline{CA} \parallel \overline{FD}$;
$\overline{AD} \parallel \overline{BE} \parallel \overline{CF}$;
$\overline{AB}$ and $\overline{CF}$, $\overline{BC}$ and $\overline{AD}$, $\overline{CA}$ and $\overline{BE}$,
$\overline{AB}$ and $\overline{FD}$, $\overline{AB}$ and $\overline{FE}$ etc. are pairs of skew lines.

15.2 A Line Perpendicular to a Plane

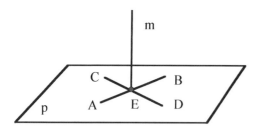

(1) If a line is not in a plane, it intersects a plane in exactly one point.
    e.g. line m intersects plane p at point E

(2) If a line is perpendicular to a plane, it is perpendicular to each line in the plane through the point of intersection.
    e.g. $m \perp \overleftrightarrow{AB}$, $m \perp \overleftrightarrow{CD}$

(3) Through a given point (on the plane or not on the plane), there is one and only one line perpendicular to the given plane.

(4) Two lines perpendicular to the same plane are parallel and coplanar.

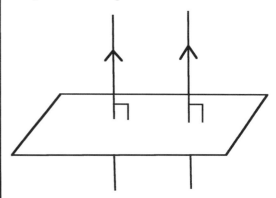

15.3 Dihedral Angle

A **dihedral angle** is the angle formed by two intersecting planes.

The measure of a dihedral angle:

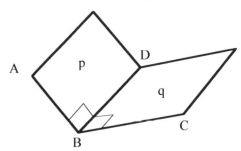

$\overleftrightarrow{BD}$ is the intersecting line of plane p and plane q.
Let both $\overrightarrow{BA}$ and $\overrightarrow{BC}$ be perpendicular to $\overleftrightarrow{BD}$.
$\angle ABC$ is the measure of the dihedral angle.

15.4 Perpendicular Planes

(1) Two perpendicular planes intersect to form a right dihedral angle.

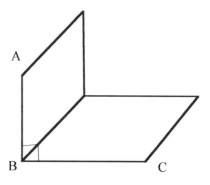

32.

(2) If plane a and plane b are both perpendicular to plane c, then their line of intersection m is perpendicular to plane c.

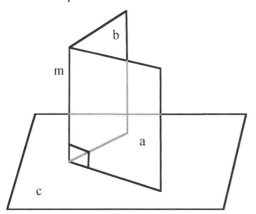

(3) If a plane contains a line perpendicular to another plane, then these two planes are perpendicular.

15.5 Parallel Planes

(1) Parallel planes are equidistant everywhere.

(2) If two planes are perpendicular to a same line, then they are parallel.

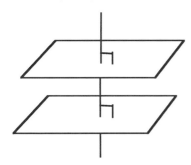

(3) If a plane intersects two parallel planes, then the intersection is two parallel lines.

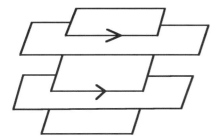

16. GEOMETRIC MEASUREMENTS

16.1 General Rules

Union of Two Regions:
The area of the union of two regions is the sum of the areas minus the area of the intersection.

$$\text{Area}(A \cup B) = \text{Area } A + \text{Area } B - \text{Area}(A \cap B)$$

e.g. Find the area of the following figure.

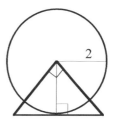

Area O + Area △ - Area of Intersection
$$= \pi \cdot r^2 + \frac{1}{2} \cdot b \cdot h - \frac{1}{4} \cdot \pi \cdot r^2$$
$$= \pi \cdot 2^2 + \frac{1}{2} \cdot 4 \cdot 2 - \frac{1}{4} \cdot \pi \cdot 2^2 = 3\pi + 4$$

Similar Objects
Similar objects have a dimension ratio r, their area ratio is r^2 and their volume ratio is r^3.

Cavalieri's Principle
Solids with equal altitudes and equal cross section areas at each height have the same volume.

e.g.

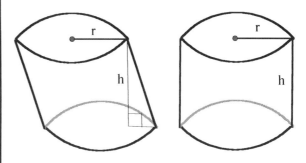

The oblique cylinder and the right cylinder have equal altitude and equal base area. They have equal volume.

II. Basic Geometry

16.2 Formulas for Measuring in Two Dimensions

1 yd = 3 ft,   1 ft = 12 in,   1 mile = 5280 ft
1 m = 100 cm,   1 m = 1000 mm

(1) Circle

Circumference $C = 2\pi r = \pi d$ r: radius, d: diameter
Area $A = \pi r^2$

e.g.
When r is doubled, C is doubled and A increases 4 times.

(2) Square

Perimeter $P = 4s$ s: length of the side
Area $A = s^2$

(3) Rectangle

Perimeter $P = 2l + 2w$ l: length   w: width
Area $A = l \cdot w$

(4) Parallelogram

Perimeter P = sum of the 4 sides   b: base, h: height
Area $A = b \cdot h$

(5) Trapezoid

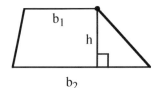

P = sum of 4 sides

$A = \dfrac{b_1 + b_2}{2} \cdot h$

(6) Rhombus

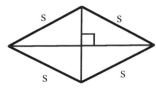

$P = 4s$

$A = \dfrac{1}{2} \cdot d_1 \cdot d_2$

d_1 and d_2 are diagonals

(7) Triangle

P = sum of 3 sides

$A = \dfrac{1}{2} \cdot b \cdot h$

16.3 Formulas for Measuring in Three Dimensions

(8) Right Prism

Volume $V = Bh$ B: Base Area, h: height
Lateral Area = Sum of all side areas
Surface Area = 2 Base Areas + Lateral Area

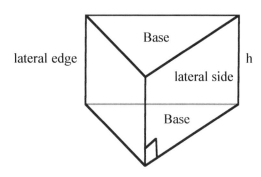

All of the lateral edges are congruent and parallel.

Rectangular Prism
Volume $V = lwh$ l: length, w: width, h: height
Surface Area $SA = 2wl + 2wh + 2hl$

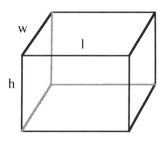

Cube
Volume $V = s^3$ s : length of the side
Surface Area $SA = 6s^2$

II. Basic Geometry

(9) Right Circular Cylinder

Volume $V = Bh$ B: area of the circular base πr^2
      h: height
Lateral Area $L = 2\pi rh$

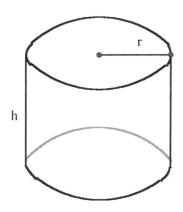

(10) Pyramid

Volume $V = \dfrac{1}{3}Bh$ B: Base Area, h: height

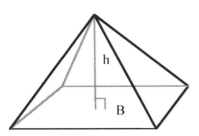

The base of a regular pyramid is a regular polygon. The lateral sides of a regular pyramid are congruent isosceles triangles.

(11) Right Circular Cone

Volume $V = \dfrac{1}{3}Bh$ B: Base Area = πr^2, h: height
Lateral Area $L = \pi rl$ l: slant height

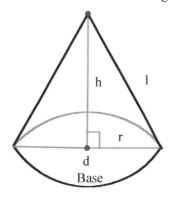

e.g.
A right circular cone has a diameter of 16 and a height of 18. Find the volume.
(a) Express the answer in terms of π.
(b) Express the answer to the nearest tenth.

$r = \dfrac{d}{2} = \dfrac{16}{2} = 8$

$B = \pi r^2 = \pi \cdot 8^2 = 64\pi$

$V = \dfrac{1}{3}Bh = \dfrac{1}{3} \cdot 64\pi \cdot 18$

(a). $V = 384\pi$

(b). $V = 1206.4$

Frustum of Right Circular Cone

A **frustum** is a portion of a cone between two parallel planes.

Volume: $V = \dfrac{1}{3}\pi h(r_1^2 + r_1 r_2 + r_2^2)$

Lateral Area: $L = \pi(r_1 + r_2) \cdot l$

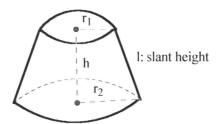

l: slant height

II. Basic Geometry

(12) Sphere

Volume $\quad V = \dfrac{4}{3}\pi r^3$

Surface Area $\quad SA = 4\pi r^2$

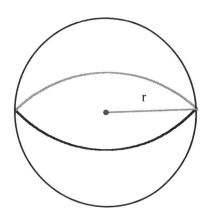

Frustum of Sphere

Lateral Area: $L = 2\pi Rh$

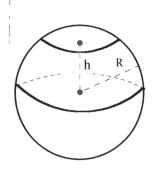

e.g.
A sphere has a surface area of 100π in^2. Find the volume of the sphere, to the nearest cubic inch.

$SA = 4\pi r^2 = 100\pi$
$r = 5$

$V = \dfrac{4}{3}\pi r^3 = \dfrac{500\pi}{3} \approx 524$ in^3

16.4 Error in Measurement

Absolute Error = |Measured Value - Actual Value|

Relative Error = $\dfrac{\text{Absolute Error}}{\text{Actual Value}}$

Percent of Error = Relative Error × 100%

e.g.
Actual value of the side of a cube is 10.0 cm. Measured value is 10.5 cm. Find the relative error and percent of error in the surface area.

Relative Error = $\dfrac{|6(10.5)^2 - 6(10)^2|}{6(10)^2} = 0.1025$

Percent of Error = 0.1025 × 100% = 10.25%

III. Coordinate Geometry and Functions

17. COORDINATE GEOMETRY

17.1 Coordinate Plane

The horizontal line is x-axis and the vertical line is y-axis. Their point of intersection, O, is the origin. Coordinate Plane has four Quadrants I, II, III, and IV.

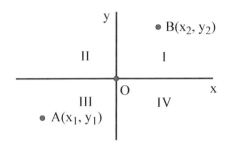

Every point in the plane corresponds to an ordered pair of numbers (x,y), and vice versa.

17.2 Slope, Midpoint, Distance, and Centroid

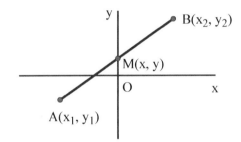

slope $\quad m = \dfrac{y_2 - y_1}{x_2 - x_1}$

midpoint $\quad M(\bar{x}, \bar{y}) = M\left(\dfrac{x_1 + x_2}{2}, \dfrac{y_1 + y_2}{2}\right)$

distance $\quad d = \sqrt{(x_2 - x_1)^2 + (y_2 - y_1)^2}$

e.g. $\overline{AB}$ has midpoint M(1,4) and one end B(3,5). Find the coordinates of the other end A.

$$1 = \dfrac{x_1 + 3}{2}, \quad 4 = \dfrac{y_1 + 5}{2}$$

Solve for x_1 and y_1. $A(x_1, y_1) = A(-1, 3)$

e.g. The centroid of a triangle is
$$\left(\dfrac{x_1 + x_2 + x_3}{3}, \dfrac{y_1 + y_2 + y_3}{3}\right)$$

17.3 Coordinate Geometric Proofs

(1) To prove a parallelogram:
Method 1: (Slope formula)
Two pairs of opposite sides are parallel - the same slope.
Method 2: (Midpoint formula)
Diagonals have the same midpoint - bisect each other.

(2) To prove a rhombus:
(Distance formula)
4 sides have the same length.

(3) To prove a rectangle:
(Slope formula)
Opposite sides are parallel, adjacent sides are perpendicular.

(4) To prove a trapezoid:
(Slope formula)
One pair of the opposite sides are parallel - the same slope, and the other pair of the opposite sides are not parallel - different slopes.

e.g. The quadrilateral ABCD has vertices A(-5, -2), B(-5, 3), C(4, 6), and D(7, 2). Prove by coordinate geometry that quadrilateral ABCD is an isosceles trapezoid.

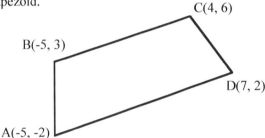

Prove:

Slope of $\overline{AD} = \dfrac{2 - (-2)}{7 - (-5)} = \dfrac{4}{12} = \dfrac{1}{3}$

Slope of $\overline{BC} = \dfrac{6 - 3}{4 - (-5)} = \dfrac{3}{9} = \dfrac{1}{3}$

Slope of $\overline{AD}$ = Slope of $\overline{BC}$ $\quad \overline{AD} \parallel \overline{BC}$

Slope of $\overline{AB} = \dfrac{3 - (-2)}{-5 - (-5)} = \dfrac{5}{0}$ (vertical line)

Slope of $\overline{CD} = \dfrac{2 - 6}{7 - 4} = \dfrac{-4}{3}$

Slope of $\overline{AB} \neq$ Slope of $\overline{CD}$ $\quad \overline{AB}$ is not $\parallel \overline{CD}$
Therefore ABCD is a trapezoid.

$AB = \sqrt{[-5 - (-5)]^2 + [3 - (-2)]^2} = 5$
$CD = \sqrt{(7 - 4)^2 + (2 - 6)^2} = 5$
$AB = CD$
Therefore ABCD is an isosceles trapezoid.

III. Coordinate Geometry and Functions

18. RELATIONS AND FUNCTIONS

18.1 Relations and Functions

A **relation** is a set of ordered pairs of data.

The **domain** is the set of all first elements of the ordered pairs.

The **range** is the set of all second elements of the ordered pairs.

A **function** is a special relation in which each element of the domain corresponds to one and only one element in the range. That means the first element in the ordered pairs cannot repeat in a function.

A function is a rule that assigns each element of the domain to exactly one element in the range.

e.g. $\{(2, 1), (3, 1), (4, 3), (5, 4)\}$ is a function.

$\{(1, 2), (1, 3), (3, 4), (4, 5)\}$ is not a function, since $(1, 2)$ and $(1, 3)$ have the same first element "1".

e.g.

Relation (Not a Function)

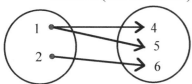

Function

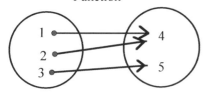

One-to-One Function

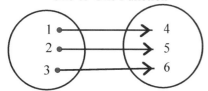

Vertical Line Test:

If any vertical line intersects the graph at only one point, then the relation is a function.

e.g. $y^2 = x$ is equivalent to $y = \pm\sqrt{x}$.
It is not a function.

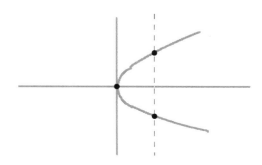

Horizontal Line Test:

If any horizontal line intersects the graph at only one point, then the function is a **one-to-one function**.

e.g.
$y = x^2$ is a function, but not a one-to-one function.

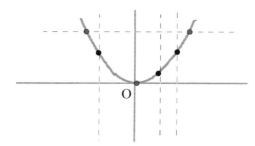

If we restrict the domain of $y = x^2$ to $x \geq 0$, then the function is a one-to-one funcion.

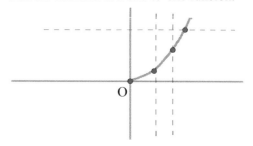

III. Coordinate Geometry and Functions

Domain and Range

e.g. $y = x^2$ Domain: $\{x \mid x \text{ all real numbers}\}$
  Range: $\{y \mid y \geq 0\}$

e.g. $y = \sqrt{x}$ Domain: $\{x \mid x \geq 0\}$
  Range: $\{y \mid y \geq 0\}$

e.g. $y = \dfrac{1}{x^2 - 9}$ Domain: $\{x \mid x \text{ all reals except } \pm 3\}$

e.g. $y = \dfrac{1}{\sqrt{x-3}}$ Domain: $\{x \mid x > 3\}$

Interval Notation

e.g.
$(2, 5)$ represents $\{x \mid 2 < x < 5\}$
$[2, 5]$ represents $\{x \mid 2 \leq x \leq 5\}$
$(2, 5]$ represents $\{x \mid 2 < x \leq 5\}$
$[2, 5)$ represents $\{x \mid 2 \leq x < 5\}$

e.g.
$(-\infty, \infty)$ represents $\{x \mid x \text{ all real numbers}\}$
$(-\infty, -5)$ represents $\{x \mid x < -5\}$
$[5, \infty)$ represents $\{x \mid x \geq 5\}$
$(-\infty, -5) \cup [5, \infty)$ represents $\{x < -5 \text{ or } x \geq 5\}$

Function Notation

e.g. If $f(x) = \dfrac{2x}{x-1}$

then $f(5) = \dfrac{2 \cdot 5}{5-1} = \dfrac{10}{4} = \dfrac{5}{2}$

$f(a+2) = \dfrac{2(a+2)}{(a+2)-1} = \dfrac{2a+4}{a+1}$

$f(x^2) = \dfrac{2(x^2)}{(x^2)-1} = \dfrac{2x^2}{x^2-1}$

If $g(x) = x^2 - 5x + 6$

then $g(3) = (3)^2 - 5(3) + 6 = 0$

$g(2x) = (2x)^2 - 5(2x) + 6$
$= 4x^2 - 10x + 6$

Set Notation

e.g. $\{(x, 2x^2 - 5) \mid x \text{ real}\}$
$\{(x, \sqrt{x} - 5) \mid x \text{ real}, x \geq 0\}$
$\{(x, \dfrac{x}{x-5}) \mid x \text{ real}, x \neq 5\}$
$\{(x, \sqrt{25 - x^2}) \mid x \text{ real}, -5 \leq x \leq 5\}$

18.2 Increasing, Decreasing, and Constant Functions

When $a < b$ for any a and b in the interval
$f(a) < f(b)$ Increasing function
$f(a) > f(b)$ Decreasing function
$f(a) = f(b)$ Constant function

e.g. $y = x^2$, Increasing on $(0, \infty)$ and decreasing on $(-\infty, 0)$.

18.3 Piecewise Defined Functions

If a function is defined by different rules in its domain, it is called **piecewise defined function**.

e.g. $f(x) = \begin{cases} -2x - 1 & \text{if } x \leq -2 \quad \text{decreasing} \\ 3 & \text{if } -2 < x < 1 \quad \text{constant} \\ 2x + 1 & \text{if } x \geq 1 \quad \text{increasing} \end{cases}$

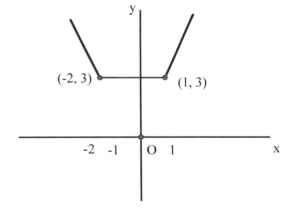

III. Coordinate Geometry and Functions

18.4 Odd and Even Functions

Odd Function: $f(-x) = -f(x)$

An odd function is symmetric about the origin.

e.g. $f(x) = 2x^3$
$f(-x) = 2(-x)^3 = -2x^3 = -f(x)$

Even Function: $f(-x) = f(x)$

An even function is symmetric about the y-axis.

e.g. $f(x) = 3x^2 + 5$
$f(-x) = 3(-x)^2 + 5 = 3x^2 + 5 = f(x)$

A constant is an even function.
The sum of odd functions is an odd function.
The sum of even functions is an even function.
The product of two odd functions is an even function.
The product of two even functions is an even function.
The product of an odd function and an even function is an odd function.

18.5 Maximum and Minimum

A value $f(c)$ is a **local maximum (relative maximum)** of $f(x)$ if there is an open interval (a, b) containing a value c such that $f(x) \leq f(c)$ for all values of x in (a, b).

A value $f(d)$ is a **local minimum (relative minimum)** of $f(x)$ if there is an open interval (a, b) containing a value d such that $f(x) \geq f(d)$ for all values of x in (a, b).

The **absolute maximum** is the largest value of $f(x)$ in its domain.

The **absolute minimum** is the smallest value of $f(x)$ in its domain.

e.g. The vertex of a parabola is the absolute maximum if it opens downward.
The vertex of a parabola is the absolute minimum if it opens upward.

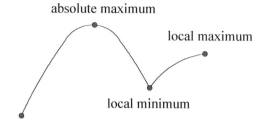

18.6 Average Rate of Change

The average rate of change over the interval $[a, b]$ is

$$\frac{f(b) - f(a)}{b - a}$$

e.g. A particle moves along a straight line with the position function $d(t)$. The **average velocity** over the time period $(t_2 - t_1)$ is

$$\bar{v} = \frac{d(t_2) - d(t_1)}{t_2 - t_1}$$

e.g. **Average Speed:** $s = \dfrac{d}{t}$ d: total distance, t: time

e.g. A car travels 300 miles in 5 hours, the average speed

$$s = \frac{300}{5} = 60 \text{ miles/hr.}$$

e.g. Tom drove 120 miles to his friend's house and the same distance back home. It took him 2 hours to drive there and 3 hours to drive back.

The average speed of driving out:
$$s_1 = \frac{120}{2} = 60 \text{ miles/hr.}$$
The average speed of driving back:
$$s_2 = \frac{120}{3} = 40 \text{ miles/hr.}$$
The average speed of the whole trip:
$$s = \frac{120 + 120}{3 + 2} = \frac{240}{5} = 48 \text{ miles/hr.}$$

(Note: $s \neq \dfrac{s_1 + s_2}{2} = 50 \text{ miles/hr.}$)

III. Coordinate Geometry and Functions

18.7 Composition of Functions

The composition of functions $y_1 = f(x)$ and $y_2 = g(x)$ denoted $f \circ g$ is the function:

$$(f \circ g)(x) = f(g(x))$$

The domain of $f \circ g$ is the set of all x-values in the domain of g such that $g(x)$ is in the domain of f.

e.g. $f(x) = x + 1$ and $g(x) = \sqrt{x}$
$(f \circ g)(x) = f(g(x)) = f(\sqrt{x}) = \sqrt{x} + 1$
Domain: $[0, \infty)$

$(g \circ f)(x) = g(f(x)) = g(x + 1) = \sqrt{x+1}$
Domain: $[-1, \infty)$

Note: $(f \circ g)(x) \neq (g \circ f)(x)$

e.g. $f(x) = x^2 - 1$ and $g(x) = \sqrt{x}$
$(f \circ g)(x) = f(g(x)) = f(\sqrt{x}) = x - 1$
Domain: $[0, \infty)$

Note: Domain is not $(-\infty, \infty)$ because $g(x) = \sqrt{x}$.

$(g \circ f)(x) = g(f(x)) = g(x^2 - 1) = \sqrt{x^2 - 1}$
$x^2 - 1 \geq 0$
Domain: $(-\infty, -1] \cup [1, \infty)$

e.g. $f(x) = x^2 - 1$, $g(x) = x + 1$

$(f \circ g)(x) = f(g(x)) = f(x + 1)$
$= (x + 1)^2 - 1 = x^2 + 2x$
$(f \circ g)(2) = f(g(2)) = f(2 + 1)$
$= f(3) = 3^2 - 1 = 8$
$(g \circ f)(x) = g(f(x)) = g(x^2 - 1)$
$= (x^2 - 1) + 1 = x^2$
$(g \circ f)(2) = g(f(2)) = g(2^2 - 1)$
$= g(3) = 3 + 1 = 4$

e.g. $f(x) = \sin x$ and $g(x) = x^2$
$(f \circ g)(x) = f(g(x)) = f(x^2) = \sin(x^2)$
$(g \circ f)(x) = g(f(x)) = g(\sin x) = (\sin x)^2 = \sin^2 x$

18.8 Inverse Functions

A function g is the inverse of the function f if
$(f \circ g)(x) = x$ for each x in the domain of g, and
$(g \circ f)(x) = x$ for each x in the domain of f.
The inverse function $g(x)$ is denoted by $f^{-1}(x)$.
$(f \circ f^{-1})(x) = x$ and $(f^{-1} \circ f)(x) = x$

Note: $f^{-1}(x)$ is not $\dfrac{1}{f(x)}$.

The domain of the inverse function is the range of the original function.

e.g. Original $f(x) = \{(1, 1), (2, 4), (3, 9)\}$
Domain: $\{x \mid x = 1, 2, 3\}$
Range: $\{y \mid y = 1, 4, 9\}$

Inverse $f^{-1}(x) = \{(1, 1), (4, 2), (9, 3)\}$
Domain: $\{x \mid x = 1, 4, 9\}$
Range: $\{y \mid y = 1, 2, 3\}$

For every one-to-one function $f(x)$, there is an inverse function $f^{-1}(x)$. (passing both vertical and horizontal line tests)

e.g. $f(x) = x^2$

It is a function. (passing the vertical line test)
But it is not a one-to-one function. (fails the horizonal line test)
With **restricted domain** $[0, \infty)$, $f(x) = x^2$ is a one-to-one fuction. Its inverse is $f^{-1}(x) = \sqrt{x}$.

Find the Inverse:

e.g. $f(x) = 3x + 5$ find $f^{-1}(x)$
$y = 3x + 5$ express $f^{-1}(x)$ as y
$x = 3y + 5$ interchange x and y
$y = \dfrac{x - 5}{3}$ solve for y in terms of x
$f^{-1}(x) = \dfrac{x - 5}{3}$

The graph of $f^{-1}(x)$ is the reflection of $f(x)$ in line $y = x$.

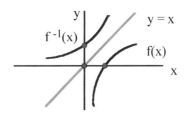

III. Coordinate Geometry and Functions

18.9 Functions Under a Transformation

Translation:

$y = f(x)$ _____ $T_{a,b}$ _____ $y = f(x - a) + b$

$y = f(x)$ translates a units right and b units up.
(Assume a, b > 0)

Reflection:

$y = f(x)$ _____ r x-axis _____ $y = -f(x)$

$y = f(x)$ reflects about the x-axis.

$y = f(x)$ _____ r y-axis _____ $y = f(-x)$

$y = f(x)$ reflects about the y-axis.

$y = f(x)$ _____ r y=x _____ $y = f^{-1}(x)$

$y = f(x)$ reflects about the y = x line.

Dilation:

$y = f(x)$ _____ vertical stretch if k > 1 _____ $y = kf(x)$
vertical shrink if 0 < k < 1

$y = f(x)$ _____ horizontal shrink if k > 1 _____ $y = f(kx)$
horizontal stretch if 0 < k < 1

The transformation rules for functions are different from the transformation rules for images.

Use the graphing calculator to verify the answer.

e.g. Inverse function reflects about y = x

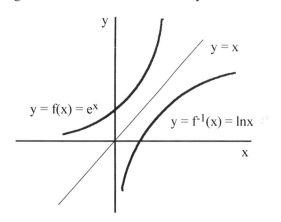

e.g. $y = x^2$ _____ $T_{5,2}$ _____ $y = (x-5)^2 + 2$
$y = x^2$ _____ r x-axis _____ $y = -x^2$

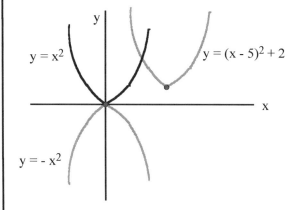

e.g.

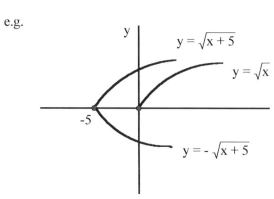

$y = \sqrt{x}$ _____ moved 5 units to the left _____ $y = \sqrt{x+5}$

$y = \sqrt{x+5}$ _____ reflected in the x-axis _____ $y = -\sqrt{x+5}$

e.g.

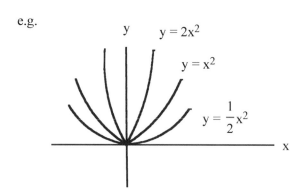

$y = x^2$ _____ stretched vertically by a factor of 2 _____ $y = 2x^2$

$y = x^2$ _____ shrunk vertically by a factor of 1/2 _____ $y = \frac{1}{2}x^2$

III. Coordinate Geometry and Functions

19. IMPORTANT FUNCTIONS AND RELATIONS

19.1 Direct Variation

A straight line passing through the origin

$$y = mx \quad \text{or} \quad \frac{y}{x} = m$$

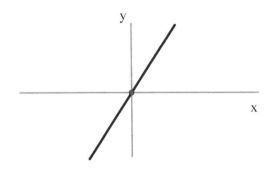

To solve a problem, use $\dfrac{x_1}{x_2} = \dfrac{y_1}{y_2}$

e.g. The distance varies directly with the time that a car travels at a constant speed.

$$d = s \cdot t \quad (\text{s is a constant})$$

19.2 Inverse Variation (Function)

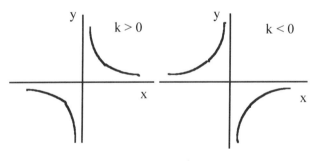

$$xy = k \quad \text{or} \quad y = \frac{k}{x}$$

To solve a problem, use $x_1 \cdot y_1 = x_2 \cdot y_2$

e.g. The speed varies inversely to the time when a car travels over a certain distance.

$$s \cdot t = d \quad (\text{d is a constant})$$

19.3 Linear Function (First Degree)

A straight line can be represented as a linear function; the graph of a linear function is a straight line.

Slope-Intercept Form: $y = mx + b$
where m is the slope and b is the y-intercept.

The slope m is considered the **rate of change**.

Point-Slope Form: $y - y_1 = m(x - x_1)$
the line passes through the point (x_1, y_1) and has slope m.

Intercept Form: $\dfrac{x}{a} + \dfrac{y}{b} = 1$
with x-intercept a and y-intercept b

General Form: $Ax + By + C = 0$

e.g.

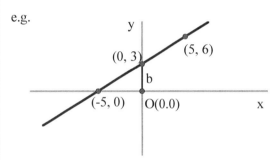

$b = 3, \quad m = \dfrac{3 - 0}{0 - (-5)} = \dfrac{3}{5}$

Slope-Intercept Form: $y = \dfrac{3}{5}x + 3$

Point Slope Form: $y - 3 = \dfrac{3}{5}x \quad$ for point (0, 3)

or $\quad y = \dfrac{3}{5}(x + 5) \quad$ for point (-5, 0)

or $\quad y - 6 = \dfrac{3}{5}(x - 5) \quad$ for point (5, 6)

Intercept Form: $-\dfrac{x}{5} + \dfrac{y}{3} = 1$

Rewrite $\quad y = \dfrac{3}{5}x + 3$

$\quad 5y = 3x + 15$

General Form: $3x - 5y + 15 = 0$

III. Coordinate Geometry and Functions

(1) The Slopes of Lines:

Two parallel lines have the same slope ($m_1 = m_2$).

Two perpendicular lines: $m_2 = -\dfrac{1}{m_1}$ or $m_1 \cdot m_2 = -1$

The slope of a horizontal line is zero ($m = 0$).
The slope of a vertical line is undefined.

Special cases:

Direct Variation: when $b = 0$, $y = mx$
Vertical line: $x = a$
Horizontal line: $y = b$

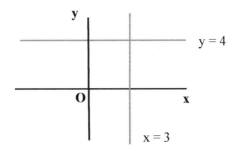

e.g. Find the slope and y-intercept of $3x - 2y = 12$.

Write the equation in slope and y-intercept form:
$y = \dfrac{3}{2}x - 6$, slope $m = \dfrac{3}{2}$ and y-intercept $b = -6$

e.g. Write the equation of a line through $(3, -2)$ and $(6, 4)$.

First find the slope $m = \dfrac{4 - (-2)}{6 - 3} = \dfrac{6}{3} = 2$

$y = 2x + b$, replace x by 6 and y by 4
$4 = 2 \cdot 6 + b$ solve for $b = -8$
We have the equation of the line $y = 2x - 8$

e.g. Write the equation of a line passing through the origin and perpendicular to the line $y = 2x + 3$.

Since the line passes through the origin: $y = mx$ ($b = 0$).

$m = -\dfrac{1}{m_1} = -\dfrac{1}{2}$ (m_1 is the slope of $y = 2x + 3$)

Solution: $y = -\dfrac{1}{2}x$

(2) The Distance from a Point to a Line:

e.g. Find the distance from point $P(1, 5)$ to the line l_1: $y = x - 2$

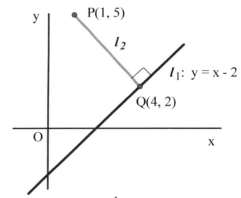

$m_1 = 1$, $m_2 = \dfrac{1}{m_1} = -1$

The equation of l_2: $y = m_2 x + b = -x + b$
Plug in $(1, 5)$ to find the value of b:
$$5 = -1 + b$$
$$b = 6$$
$$y = -x + 6$$
Find the intersecting point of l_1 and l_2:
$$y = x - 2 \quad (1)$$
$$y = -x + 6 \quad (2)$$
Add Eq.(1) and Eq.(2):
$$2y = 4, \quad y = 2$$
Replace $y = 2$ in Eq.(1):
$$x = 4$$
The intersecting point $Q(4, 2)$.
The distance from the point to the line:
$$d = \sqrt{(x_2 - x_1)^2 + (y_2 - y_1)^2}$$
$$= \sqrt{(4 - 1)^2 + (2 - 5)^2} = 3\sqrt{2}$$

The General Distance Formula from point $P(x_0, y_0)$ to the line $Ax + By + C = 0$

$$d = \dfrac{|Ax_0 + By_0 + C|}{\sqrt{A^2 + B^2}}$$

For the above example:
Point $P(1, 5)$ and Line: $x - y - 2 = 0$

$$d = \dfrac{|(1)(1) + (-1)(5) + (-2)|}{\sqrt{(1)^2 + (-1)^2}} = \dfrac{6}{\sqrt{2}} = 3\sqrt{2}$$

III. Coordinate Geometry and Functions

19.4 Greatest Integer Function

$y = [x]$

Domain: $\{x \mid x \text{ all real numbers}\}$

Range: $\{y \mid y \text{ all integers}\}$

e.g. $y = [1.6] = 1$
      $y = [-1.6] = -2$

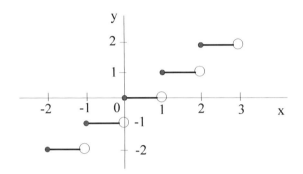

19.5 Linear Inequalities

e.g. The solution of $y < x + 2$ is the region under $y = x + 2$;

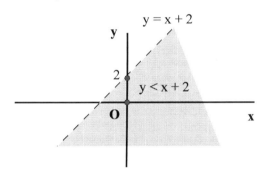

e.g. The solution of $y \geq x + 2$ is the region above $y = x + 2$ and including the line $y = x + 2$

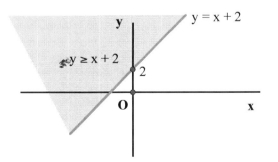

Rectangular Region

e.g. $\{(x, y) \mid -2 \leq x \leq 5, 3 \leq y \leq 8\}$

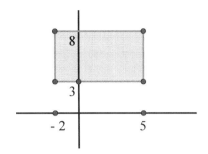

Triangular Region

e.g. $\{(x, y) \mid y \geq 0, y \leq x, y \leq -x + 8\}$

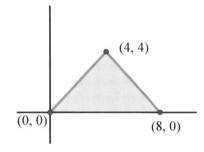

19.6 Absolute Value Functions

$y = |x|$

when $x < 0$
$y = -x$

when $x \geq 0$,
$y = x$

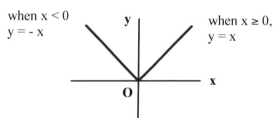

e.g. $y = |2x - 4|$

when $2x - 4 < 0$
$x < 2$
$y = -(2x - 4)$
$y = -2x + 4$

when $2x - 4 \geq 0$
$x \geq 2$
$y = 2x - 4$

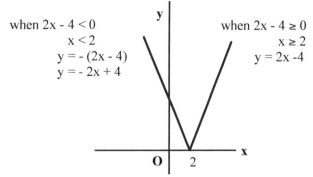

III. Coordinate Geometry and Functions

19.7 Quadratic Functions and Parabolas

General Form:

$$y = f(x) = ax^2 + bx + c \quad \text{where } a \neq 0$$

(1). Axis of Symmetry: $x = -\dfrac{b}{2a}$

or $\quad x = \dfrac{x_1 + x_2}{2}$

if the function has two real roots x_1 and x_2.

(2). Vertex (Turning Point): (x, y)

$$x = -\dfrac{b}{2a}, \quad y = f(x) = f\left(-\dfrac{b}{2a}\right)$$

(3). Opening:

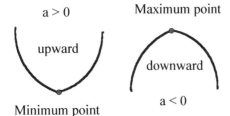

e.g. $y = 12x - 2x^2$
$y = -2x^2 + 12x$ write in standard form
$a = -2, \; b = 12, \; c = 0$

(1). Axis of Symmetry: $x = -\dfrac{b}{2a} = -\dfrac{12}{2(-2)} = 3$

(2). Vertex: $x = 3, \; y = -2(3)^2 + 12(3) = 18$
$(3, 18)$

(3). Opening: $a = -2 < 0$
It has a maximum of 18 at $x = 3$.

Vertex Form:

$$y = f(x) = a(x - h)^2 + k$$

where (h, k) is the vertex of the parabola.

Factored Form:

$$y = f(x) = a(x - x_1)(x - x_2)$$

where x_1 and x_2 are the two x-intercepts (real roots).

e.g. Sketch the graph of
$$f(x) = \dfrac{1}{2}(x + 3)(x - 5)$$

$a = \dfrac{1}{2} > 0$, the opening is upward.

$x_1 = -3$ and $x_2 = 5$
the axis of symmetry:
$$x = \dfrac{x_1 + x_2}{2} = \dfrac{-3 + 5}{2} = 1$$

$f(1) = \dfrac{1}{2}(1 + 3)(1 - 5) = -8$

Vertex: $(1, -8)$

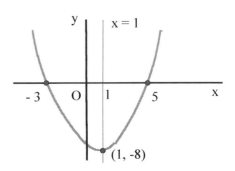

e.g. $y = x^2 - 2x - 3$
$a = 1, \; b = -2, \; c = -3$

(1). Axis of Symmetry: $x = -\dfrac{b}{2a} = \dfrac{-(-2)}{2(1)} = 1$

(2). Vertex: $x = 1, \; y = (1)^2 - 2(1) - 3 = -4$
Vertex: $(1, -4)$

(3). Opening: $a = 1 > 0$, upward
It has a minimum of -4 at $x = 1$

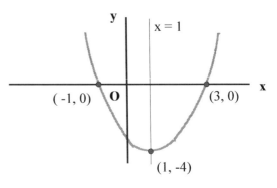

(4). Find the roots of the equation:
$x^2 - 2x - 3 = 0$
$(x - 3)(x + 1) = 0$
$x = -1, \quad x = 3$
x-intercepts $(-1, 0)$ and $(3, 0)$

III. Coordinate Geometry and Functions

Projectile Motion (Vertical Launch)

$$h = -\frac{1}{2}gt^2 + v_0 t + h_0$$

where h is the height at time t.
    g is the gravitational acceleration.
    g = 32 ft/sec² or 9.8 m/sec²
    v_0 is the initial vertical velocity, positive for upward and negative for downward, zero for a free fall.
    h_0 is the initial height.

e.g. A ball is thrown upward from the ground level with an initial velocity of 48 ft/sec.
(1) In how many seconds does the ball reach the ground?
(2) What is the maximum height that the ball can reach?

Use the formula:
$$h = -\frac{1}{2}gt^2 + v_0 t + h_0$$
    where g = 32, v_0 = 48, h_0 = 0
$$h = -16t^2 + 48t$$

(1) When the ball reaches the ground, h = 0.
    $0 = -16t^2 + 48t$
    $t(16t - 48) = 0$
    t = 0,   t = 3
    After 3 seconds the ball reaches the ground.

(2) Find the axis of symmetry:
$$x = -\frac{b}{2a} = -\frac{48}{2 \cdot (-16)} = 1.5$$
or $$x = \frac{x_1 + x_2}{2} = \frac{0 + 3}{1} = 1.5$$

The maximum height:
    $h(1.5) = -16(1.5)^2 + 48(1.5) = 36$ ft

19.8 Quadratic Inequalities in Two Variables

e.g. $x^2 - 2x < 5 + y$

Rewrite it in the standard form:
    $y > x^2 - 2x - 5$
Graph $y = x^2 - 2x - 5$ in dashed line.

The shaded region above the curve is the solution

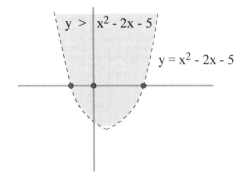

e.g. $x^2 - 2x \geq 5 + y$

Rewrite it in the standard form:
    $y \leq x^2 - 2x - 5$
Graph $y = x^2 - 2x - 5$ in solid line.

The solid line and the shaded region under the curve are the solution.

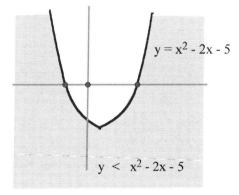

III. Coordinate Geometry and Functions

19.9 Equations of Circles (Relations, not Functions)

The center-radius equation of a circle with radius r and center (h, k)

$$(x - h)^2 + (y - k)^2 = r^2$$

Special Case:

The center of the circle is at the origin.
$$x^2 + y^2 = r^2$$

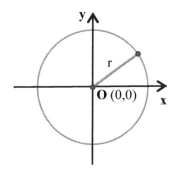

The General Equation of a Circle:

$$x^2 + y^2 + ax + by + c = 0$$

e.g. $x^2 + y^2 + 4x - 6y - 12 = 0$
Find its center and radius, and graph it.

Complete the square:
$x^2 + 4x + y^2 - 6y = 12$
$x^2 + 4x + (\frac{4}{2})^2 + y^2 - 6y + (\frac{-6}{2})^2$
$\quad = 12 + (\frac{4}{2})^2 + (\frac{-6}{2})^2$
$x^2 + 4x + 4 + y^2 - 6y + 9 = 12 + 4 + 9$
$(x + 2)^2 + (y - 3)^2 = 5^2$

Center is (- 2, 3) and the radius is 5

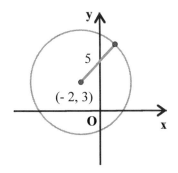

A circle has a diameter with endpoints (a, b) and (c, d). The equation of the circle can be written as:

$$(x - a)(x - c) + (y - b)(y - d) = 0$$

e.g. Write the general equation of a circle whose diameter has endpoints (-5, -2) and (3, 7).

$(x + 5)(x - 3) + (y + 2)(y - 7) = 0$
$x^2 + 2x - 15 + y^2 - 5y - 14 = 0$
$x^2 + y^2 + 2x - 5y - 29 = 0$

e.g. Find the equation of the tangent line to the circle $x^2 + y^2 = 5^2$ at the point (3, 4)

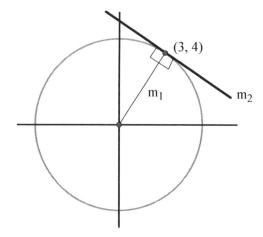

First find the slope of the **normal line** which is perpendicular to the tangent line at point (3, 4).

$$m_1 = \frac{4}{3}$$

The slope of the tangent line at point (3, 4) is

$$m_2 = -\frac{1}{m_1} = -\frac{3}{4}$$

The equation of the tangent line is

$$y - 4 = -\frac{3}{4}(x - 3) \quad \text{Point-Slope Form}$$

III. Coordinate Geometry and Functions

19.10 Exponential Functions and Equations

(1) Exponential Function:

$$y = a^x \quad \text{where } a > 0 \text{ and } a \neq 1$$

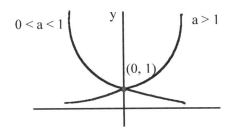

(1). Domain: $\{ x \mid x \text{ all real numbers} \}$
   Range: $\{ y \mid y > 0 \}$
(2). $a > 1$, the function is increasing (**exponential growth**);
   $a < 1$, the function is decreasing (**exponential decay**).
(3). when $x = 0$, $y = 1$ (the graphs of the exponential functions pass point $(0, 1)$).
(4). x-axis is the horizontal asymptote.
(5). $y = (\frac{1}{a})^x = a^{-x}$ and $y = a^x$ are symmetric in the y-axis.

Note:
A horizontal asymptote is a horizontal line $y = c$, such that, as x approaches infinity, y approaches but never equals c.

(2) General Form:

$$y = k \cdot a^x \quad \text{where } a > 0 \text{ and } a \neq 1$$
$$k \text{ is a constant}$$

e.g. A new car will depreciate at a rate of 8% per year. If a new car is worth $15,000, how much will it be worth after 3 years?

$$A = A_0(1 - 8\%)^t$$
$$= 15{,}000(0.92)^3$$
$$= 11{,}680$$

Answer: The car will be worth $11,680 after 3 years.

e.g. The geometric sequence is a special exponential function whose domain is the set of the counting numbers.

$$a(n) = a_1 \cdot r^{n-1} = \frac{a_1}{r} \cdot r^n = k \cdot r^n$$

(3) Exponential Models:

e.g. $A = A_0 2^{\frac{t}{96}}$
   A_0 is the original amount. (when $t = 0$)
   A is the amount at time t.

If the original amount $A_0 = 250$, find the amount A when the time is 24.
$$A = 250 \cdot 2^{\frac{24}{96}}$$
$$A = 250 \cdot 2^{0.25} = 297.3$$

e.g. $A = A_0 e^{-0.025t}$
   A_0 is the original amount;
   A is the amount at time t;

Find the **half-life** of this exponential decay.
half-life: $A = \frac{1}{2}A_0 = 0.5A_0$
$$0.5A_0 = A_0 e^{-0.025t}$$
$$0.5 = e^{-0.025t}$$
$$\ln 0.5 = -0.025t \ln e \quad (\ln e = 1)$$
$$t = \frac{\ln 0.5}{-0.025} = 27.726$$

(4) Exponential Equations:

e.g.
$$9^{x+1} = 27^x$$
$$3^{2(x+1)} = 3^{3x} \quad \text{transform to same base}$$
$$2(x+1) = 3x$$
$$x = 2 \quad \text{check: true}$$

e.g.
$$3^x = 5$$
$$\log 3^x = \log 5$$
$$x \log 3 = \log 5$$
$$x = \frac{\log 5}{\log 3} = 1.465$$

III. Coordinate Geometry and Functions

(5) Exponential Inequalities:

e.g. $9^{x+1} > 27^x$

$3^{2(x+1)} > 3^{3x}$ transform to same base

Since $f(x) = 3^x$ is an increasing function, we know $f(a) > f(b)$ if $a > b$. Therefore:

$$2(x+1) > 3x$$
$$2x + 2 > 3x$$
$$2 > x$$
$$x < 2$$

(6) Future Value and Present Value:

The **future value** of a loan is equal to the money borrowed plus the accumulated interest. The money borrowed is called the **principal** or **present value**.

Interest compounded once a year:

$$F = P(1 + r)^t$$

where F is the future value
P is the present value
r is the interest rate
t is the number of years

Interest compounded n times a year:

$$F = P(1 + \frac{r}{n})^{nt}$$

Interest compounded continuously, $n \to \infty$

$$F = P(1 + \frac{r}{n})^{nt} = P \cdot e^{rt}$$

where $e = 1 + \frac{1}{1!} + \frac{1}{2!} + \frac{1}{3!} + \cdots$
$= 2.718281828 \cdots$

(7) Interest Compounding:

Compare the money compounded annually, monthly, and continuously.
   Original investment $P = \$1000$
   Annual interest rate $r = 5\%$
   Number of years $t = 8$

(a) Interest compounded annually
$$F = P(1 + r)^t$$
$$= 1000(1 + 0.05)^8$$
$$= 1477.46$$

(b) Interest compounded monthly
$$F = P(1 + \frac{r}{n})^{nt}$$
$$= 1000(1 + \frac{0.05}{12})^{12 \cdot 8}$$
$$= 1490.59$$

(c) Interest compounded continuously
$$F = P \cdot e^{rt}$$
$$= 1000 e^{0.05 \cdot 8}$$
$$= 1491.82$$

(8) Mortgage Loan and Payments:

The future value of the loan and its accumulated interest is equal to the future value of the payments and their accumulated interest.

Assume that the interest is compounded monthly, the monthly interest rate is equal to APR/12, where APR is the **annual percentage rate**.

$$L(1+r)^t = M[1 + (1+r)^1 + (1+r)^2 + \cdots$$
$$\cdots + (1+r)^{t-1}]$$

$$L(1+r)^t = M \frac{(1+r)^t - 1}{r}$$

Where L = Loan
r = monthly interest rate = APR/12
t = number of months
M = monthly payment

e.g. Find the monthly payment for a loan of $200,000 over 30 years at APR 6%

$$r = \frac{0.06}{12} = 0.005, \quad t = 30 \cdot 12 = 360$$

$$200{,}000 \cdot 1.005^{360} = M \frac{1.005^{360} - 1}{0.005}$$

$$M = \$1199.10$$

III. Coordinate Geometry and Functions

(9) Newton's Law of Cooling:

The rate of cooling of an object is proportional to the temperature difference between the object and its surroundings. The temperature of the object is a fuction of the time:

$$T = T_s + (T_0 - T_s)e^{-kt}$$

or $\quad \dfrac{T - T_s}{T_0 - T_s} = e^{-kt}$

where T is the temperature of the object at time t,
T_s is the temperature of the surroundings,
T_0 is the initial temperature of the object,
and k is a constant.

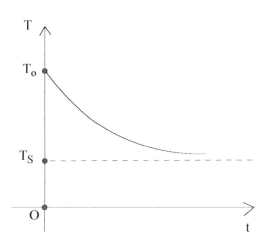

e.g. It takes 20 minutes to cool a cup of coffee from 90°C to 60°C at room temperature of 20°C.
Find the value of the constant k.

$$\dfrac{60 - 20}{90 - 20} = e^{-k \cdot 20}$$

$$\ln(\dfrac{40}{70}) = -k \cdot 20 \cdot \ln e$$

$$k = -\ln(\dfrac{40}{70}) \div 20 \approx 0.02798$$

e.g. It takes 20 minutes to cool a cup of coffee from 90°C to 60°C at room temperature of 20°C. How long does it take to cool it from 90°C to 30°C at the same room temperature?

General Solution:

$$\dfrac{T_1 - T_s}{T_0 - T_s} = e^{-kt_1}$$

$$\ln(\dfrac{T_1 - T_s}{T_0 - T_s}) = -kt_1 \qquad (1)$$

same for T_2 and t_2:

$$\ln(\dfrac{T_2 - T_s}{T_0 - T_s}) = -kt_2 \qquad (2)$$

Eq.(2) ÷ Eq.(1):

$$\dfrac{t_2}{t_1} = \dfrac{\ln(\dfrac{T_2 - T_s}{T_0 - T_s})}{\ln(\dfrac{T_1 - T_s}{T_0 - T_s})} \qquad (3)$$

Solution to this problem:
$T_0 = 90$, $T_s = 20$, $T_1 = 60$, $T_2 = 30$, $t_1 = 20$

$$t_2 = t_1 \cdot \dfrac{\ln(\dfrac{T_2 - T_s}{T_0 - T_s})}{\ln(\dfrac{T_1 - T_s}{T_0 - T_s})}$$

$$= 20 \cdot \dfrac{\ln(\dfrac{10}{70})}{\ln(\dfrac{40}{70})}$$

$$\approx 70 \text{ min.}$$

III. Coordinate Geometry and Functions

19.11 Logarithmic Functions and Equations

The logarithmic function $y = \log_a x$ is the inverse of the exponential function $y = a^x$.

$y = \log_a x$ is equivalent to $x = a^y$

e.g. $2 = \log_5 x$ is equivalent to $x = 5^2 = 25$

(1) Logarithmic Function:

$y = \log_a x$ where $a > 0$ and $a \neq 1$

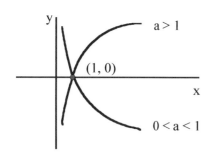

Domain: $\{ x \mid x > 0 \}$
Range: $\{ y \mid y \text{ all real numbers} \}$

$$\log_a 1 = 0, \quad \log_a a = 1, \quad \log_a \frac{1}{a} = -1$$

$$\log_a AB = \log_a A + \log_a B$$

$$\log_a \frac{A}{B} = \log_a A - \log_a B$$

$$\log_a A^n = n \cdot \log_a A, \quad \log_a \sqrt[n]{A} = \frac{1}{n} \log_a A$$

Special Cases:
$$\log_a \frac{1}{B} = -\log_a B$$
$$\log_a a^n = n$$
$$a^{\log_a n} = n$$

e.g. $\ln e^x = x, \quad e^{\ln x} = x$

e.g. $\log_2 \sqrt[3]{2} = \log_2 2^{\frac{1}{3}} = \frac{1}{3} \log_2 2 = \frac{1}{3}$

e.g. If $\log_5 x = 2$, what is the value of $\sqrt{x}$?
$\log_5 x = 2$ is equivalent to $x = 5^2 = 25$
$\sqrt{x} = \sqrt{25} = 5$

(2) Common Logarithms:

$$\log A = \log_{10} A$$

$\log(x \cdot 10^n) = n + \log x$
$0 < \log x < 1$ when $1 < x < 10$

$$10^{\log A} = A$$

e.g. $\log 100 = \log 10^2 = 2$
e.g. $\log 0.01 = \log 10^{-2} = -2$
e.g. $\log 123 = \log 1.23 \cdot 10^2 = 2 + \log 1.23$
e.g. $\log 1230 = \log 1.23 \cdot 10^3 = 3 + \log 1.23$
e.g. $\log 0.123 = \log 1.23 \cdot 10^{-1} = -1 + \log 1.23$
e.g. $\log(3.8 \times 10^{-5}) = -5 + \log 3.8$

e.g. If $\log A = 2$ and $\log B = 3$, then
$$\log \frac{\sqrt{A}}{B^3} = \frac{1}{2} \log A - 3 \log B = \frac{1}{2} \cdot 2 - 3 \cdot 3 = -8$$

(3) Natural Logarithms:

$$\ln A = \log_e A$$

$$e = \sum_{n=0}^{\infty} \frac{1}{n!} = 1 + \frac{1}{1!} + \frac{1}{2!} + \frac{1}{3!} + \cdots$$

$e = 2.718281828 \cdots$

$\ln 10 \approx 2.3 \qquad \log e \approx 0.4343$

$$e^{\ln A} = A$$

e.g. If $\ln x = 5$, then
$x = e^5 \approx 148.4$

$$\ln \frac{\sqrt{x}}{e^2} = \ln \sqrt{x} - \ln e^2$$
$$= \frac{1}{2} \ln x - 2 \ln e$$
$$= \frac{5}{2} - 2 = \frac{1}{2}$$

e.g. $e^{2\ln 5} = e^{\ln 5^2} = 5^2 = 25$

III. Coordinate Geometry and Functions

(4) Change of Base Formula

$$\log_a A = \frac{\log_b A}{\log_b a}$$

Special Case:

$$\log_a A = \frac{\log A}{\log a}$$

$$\log_a A = \frac{\ln A}{\ln a}$$

$$\log_{1/a} A = \frac{\log_a A}{\log_a \frac{1}{a}} = -\log_a A$$

e.g. $\log_2 5 = \dfrac{\log 5}{\log 2} \approx 2.3219$

or $\log_2 5 = \dfrac{\ln 5}{\ln 2} \approx 2.3219$

(5) Logarithmic Equations and Inequalities

e.g. $\log_x 4 + \log_x 9 = 2$
$\log_x 4 \cdot 9 = 2$
$x^2 = 36$
$x = 6$ ($x = -6$ rejected)

e.g. $\log_4(x^2 + 3x) - \log_4(x+5) = 1$
$\log_4 \dfrac{x^2 + 3x}{x+5} = \log_4 4$
$\dfrac{x^2 + 3x}{x+5} = 4$
$x^2 + 3x = 4x + 20$
$x^2 - x - 20 = 0$
$(x+4)(x-5) = 0$
$x = -4$ or $x = 5$ check: true

Some logarithmic equations are solved by using the exponential form.

e.g. $\log_{x+3} \dfrac{x^3 + x - 2}{x} = 2$

Rewrite the equation in exponential form:

$(x+3)^2 = \dfrac{x^3 + x - 2}{x}$

$x^2 + 6x + 9 = \dfrac{x^3 + x - 2}{x}$

$x^3 + 6x^2 + 9x = x^3 + x - 2$
$6x^2 + 8x + 2 = 0$
$3x^2 + 4x + 1 = 0$
$(3x+1)(x+1) = 0$
$3x + 1 = 0$, $x + 1 = 0$
$x = -\dfrac{1}{3}$, $x = -1$ (check: both are solutions)

e.g. $\log_{1/4}(x^2 + 3x) < \log_{1/4} 4$

Since the function $f(x) = \log_{1/4}(x)$ is decreasing, we have $f(a) < f(b)$ if $b > a$. Therefore,

$x^2 + 3x > 4$
$x^2 + 3x - 4 > 0$
Solve $x^2 + 3x - 4 = 0$
$x_1 = -4$, $x_2 = 1$
Solution: $x < -4$ or $x > 1$

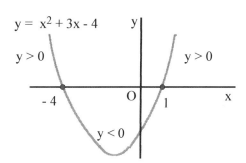

III. Coordinate Geometry and Functions

19.12 Conic Sections

(1) Ellipses (Relation)

$$\frac{x^2}{a^2} + \frac{y^2}{b^2} = 1$$

Area = πab

It is a circle when a = b

$$x^2 + y^2 = r^2 \qquad (r = a = b)$$

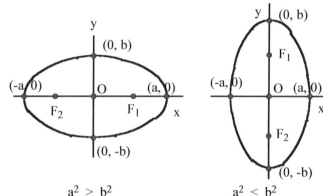

$a^2 > b^2$
$c^2 = a^2 - b^2$

$a^2 < b^2$
$c^2 = b^2 - a^2$

Foci: $F_1(c, 0)$ and $F_2(-c, 0)$
Major axis: 2a
Minor axis: 2b

Foci: $F_1(0, c)$ and $F_2(0, -c)$
Major axis: 2b
Minor axis: 2a

e.g. $4x^2 + 9y^2 = 36$
$$\frac{x^2}{9} + \frac{y^2}{4} = 1 \qquad a = 3 \quad b = 2 \quad c = \sqrt{5}$$

(2) Hyperbolas (Relation)

$$\frac{x^2}{a^2} - \frac{y^2}{b^2} = 1$$

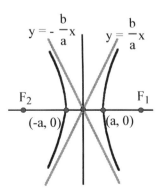

$c^2 = a^2 + b^2$
Foci: $F_1(c, 0)$ and $F_2(-c, 0)$
Asymptotes: $y = \pm \frac{b}{a}x$

(3) Parabolas

A parabola is the set of all points in a plane equidistant from a given point and a given line.

The point is called the **focus**.
The line is called the **directrix**.

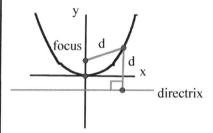

The standard forms of a parabola with the vertex (0, 0):

$y = ax^2$ Focus: $(0, \frac{1}{4a})$, Directrix: $y = -\frac{1}{4a}$

$x = ay^2$ Focus: $(\frac{1}{4a}, 0)$, Directrix: $x = -\frac{1}{4a}$

$y = ax^2$ 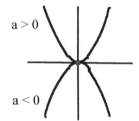 $x = ay^2$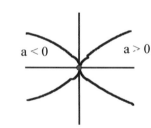

e.g.
Find the focus and the directrix for the parabola $y = -x^2$.

From the equation: $a = -1$, $\frac{1}{4a} = -\frac{1}{4}$

Focus is $(0, \frac{1}{4a}) = (0, -\frac{1}{4})$;

Directrix is $y = -\frac{1}{4a}$, $y = \frac{1}{4}$

e.g.
Find the equation in standard form for the parabola whose focus is $(-\frac{1}{2}, 0)$ and directrix is $x = \frac{1}{2}$.

$-\frac{1}{2} = \frac{1}{4a}$, $a = -\frac{1}{2}$, equation $x = -\frac{1}{2}y^2$

III. Coordinate Geometry and Functions

19.13 Polynomial Expressions and Functions

Polynomials are the most important algebraic expressions. There are many equivalent expressions and formulas.

(1) Polynomial Formulas:

$x^2 - y^2 = (x - y)(x + y)$
$x^3 - y^3 = (x - y)(x^2 + xy + y^2)$
$x^3 + y^3 = (x + y)(x^2 - xy + y^2)$
$x^4 - y^4 = (x^2 - y^2)(x^2 + y^2)$
$\quad\quad\; = (x - y)(x + y)(x^2 + y^2)$
$x^n - 1 = (x - 1)(x^{n-1} + x^{n-2} + \cdots + x^2 + x + 1)$
$x^3 - 1 = (x - 1)(x^2 + x + 1)$
$x^4 - 1 = (x - 1)(x^3 + x^2 + x + 1)$

$(a + b)^2 = a^2 + 2ab + b^2$
$(a - b)^2 = a^2 - 2ab + b^2$
$(a + b)^3 = a^3 + 3a^2b + 3ab^2 + b^3$
$(a - b)^3 = a^3 - 3a^2b + 3ab^2 - b^3$

e.g. Do the mental math:
$52 \times 48 = (50 + 2)(50 - 2)$
$\quad\quad\quad\;\; = 50^2 - 2^2$
$\quad\quad\quad\;\; = 2500 - 4 = 2496$

Since $\;\;(n + 1)^2 - n^2$
$\quad\quad = (n + 1 - n)(n + 1 + n)$
$\quad\quad = 2n + 1$,

every odd number $2n + 1$ can be expressed as the difference of the squares of two consecutive numbers.

e.g $\;\; 21 = 11^2 - 10^2 \quad\quad (n = 10)$
$\quad\quad 33 = 17^2 - 16^2 \quad\quad (n = 16)$

e.g. Show that $8^3 - 1$ is divisible by 7.
$\quad\; 8^3 - 1 = (8 - 1)(8^2 + 8 + 1)$
$\quad\quad\quad\;\; = 7(8^2 + 8 + 1)$

(2) Standard Form:

The standard form of a polynomial in one variable:

$$a_n x^n + a_{n-1} x^{n-1} + \cdots + a_2 x^2 + a_1 x + a_0$$

where n is a non-negative integer that is the degree of the polynomial; all the coefficients a_i are real numbers, and the leading coefficient $a_n \neq 0$.

(3) Polynomial Division:

$$f(x) = p(x) \cdot q(x) + r(x)$$
Dividend = (Divisor)•(Quotient) + Remainder

The long division algorithm for polynomial is the same as the one for integers.

e.g.
```
        2 4 6
    4 ) 9 8 7
        8         2 x 4
        1 8
        1 6       4 x 4
          2 7
          2 4     6 x 4
            3  (remainder)
```

e.g $\;\;(x^4 - 4x^2 + 9x - 27) \div (x^2 - x + 5)$
add the missing term as $0x^3$

```
                    x² + x - 8
      x² - x + 5 ) x⁴ + 0x³ - 4x² + 9x - 27
                   x⁴ -  x³ + 5x²
                        x³ - 9x² + 9x
                        x³ -  x² + 5x
                           - 8x² + 4x - 27
                           - 8x² + 8x - 40
                                 - 4x + 13
```

$\quad\;(x^4 - 4x^2 + 9x - 27) \quad$ Dividend
$= (x^2 - x + 5)(x^2 + x - 8) + (-4x + 13)$
$\quad\;\;$ Divisor $\quad\quad\;\;$ Quotient $\quad\;\;$ Remainder

III. Coordinate Geometry and Functions

(4) The Roots of the Function:

If c is a real number such that $f(c) = 0$, the following statements are equivalent:

1. $x = c$ is a solution or a root of the equation $f(x) = 0$.
2. $x = c$ is a zero of the function $y = f(x)$.
3. $(c, 0)$ is an x-intercept of the graph of $y = f(x)$.
4. $(x - c)$ is a factor of $f(x)$.

e.g. Find a polynomial function with given zeros 3, -4, 5.

$$f(x) = (x - 3)(x + 4)(x - 5)$$
$$= (x^2 + x - 12)(x - 5)$$
$$= x^3 - 4x^2 - 17x + 60$$

Repeated Zeros

A function may have repeated zeros.

e.g. $f(x) = (x - a)(x - b)^2(x - c)^3$
a is a zero.
b is a zero with multiplicity 2.
c is a zero with multiplicity 3.

(5) Remainder Theorem:

If a polynomial $f(x)$ is divided by $(x - c)$, then the remainder is $r = f(c)$.

e.g. Find the remainder of the following division.
$(x^3 - 2x^2 + x - 5) \div (x - 1)$
$r = f(1) = 1^3 - 2 \cdot 1^2 + 1 - 5 = -5$

(6) Factor Theorem:

A polynomial function $f(x)$ has a factor $(x - c)$ if and only if $f(c) = 0$.

e.g. Factor the following polynomial:
$x^3 - x^2 - 4x + 4$

Use graphing calculator to find all the zeros of
$f(x) = x^3 - x^2 - 4x + 4$
-2, 1, 2
$x^3 - x^2 - 4x + 4 = (x + 2)(x - 1)(x - 2)$

(7) End Behavior:

What happens to the values of $f(x)$ when x approaches positive and negative infinities.

The end behaviors are determined by the degree of the polynomial function and its leading coefficient a_n.

1. For a polynomial function of degree n which is odd:

$a_n > 0 \quad x \to \infty, \quad f(x) \to \infty$
$ \quad x \to -\infty, \quad f(x) \to -\infty$

$a_n < 0 \quad x \to \infty, \quad f(x) \to -\infty$
$\phantom{a_n < 0} \quad x \to -\infty, \quad f(x) \to \infty$

2. For a polynomial function of degree n which is even:

$a_n > 0 \quad x \to \infty, \quad f(x) \to \infty$
$ \quad x \to -\infty, \quad f(x) \to \infty$

$a_n < 0 \quad x \to \infty, \quad f(x) \to -\infty$
$\phantom{a_n < 0} \quad x \to -\infty, \quad f(x) \to -\infty$

III. Coordinate Geometry and Functions

(8) Sketch the Polynomial Function:

We can use end behaviors and zeros to help sketch the polynomial functions.

e.g. Sketch the polynomial function
$$y = (x + 3)(2x - 1)(x - 4)$$

zeros: $-3, 0.5, 4$

end behaviors: $n = 3$, $a_3 = 2 > 0$
$x \to \infty$, $y \to \infty$ going up
$x \to -\infty$, $y \to -\infty$ going down

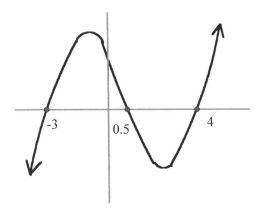

Local maxima and minima, increasing and decreasing properties are also used to help sketch the polynomial functions.

(9) Rational Zeros Theorem:

$$f(x) = a_n x^n + a_{n-1} x^{n-1} + \cdots + a_2 x^2 + a_1 x + a_0$$

where all the coefficients a_i are integers.

If $x = \dfrac{p}{q}$ is a rational zero, where p and q have no common factors, then
- p is a factor of the constant term a_0, and
- q is a factor of the leading coefficient a_n.

Rational Zeros Theorem suggests how to search for the possible rational zeros.

e.g. Solve $x^3 - 5x^2 + x + 7 = 0$
Possible zeros are $-1, 1, -7,$ and 7.
Since $f(-1) = (-1)^3 - 5(-1)^2 + (-1) + 7 = 0$,
$x = -1$ is a zero, $(x + 1)$ is a factor of $f(x)$.

We divide $f(x)$ by $(x + 1)$ to reduce the degree of the function to a quadratic function.

$$\begin{array}{r}
x^2 - 6x + 7 \\
x+1 \overline{\smash{)}\, x^3 - 5x^2 + x + 7} \\
\underline{x^3 + x^2 } \\
-6x^2 + x \\
\underline{-6x^2 - 6x } \\
7x + 7 \\
\underline{7x + 7} \\
0
\end{array}$$

Solve $x^2 - 6x + 7 = 0$

$$x = \frac{-b \pm \sqrt{b^2 - 4ac}}{2a}$$
$$= \frac{6 \pm \sqrt{36 - 28}}{2}$$
$$= 3 \pm \sqrt{2}$$

Solutions: $-1, 3 - \sqrt{2}, 3 + \sqrt{2}$

III. Coordinate Geometry and Functions

20. COMPLEX MUNBERS

20.1 Complex Numbers

The Imaginary Number i :

$i = \sqrt{-1}$, $\quad i^2 = -1$

$i^0 = 1,\quad i^1 = i,\quad i^2 = -1,\quad i^3 = -i$
$i^{4n} = 1,\quad i^{4n+1} = i,\quad i^{4n+2} = -1,\quad i^{4n+3} = -i$

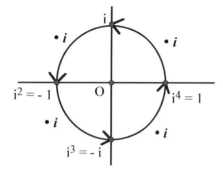

e.g. $i^{82} = i^{4 \cdot 20 + 2} = i^2 = -1$

e.g. $\sqrt{-16} + \sqrt{-9} = 4i + 3i = 7i$

e.g. $\sqrt{-16} \cdot \sqrt{-9} = 4i \cdot 3i = 12\,i^2 = -12$

(Note: $\sqrt{-16} \cdot \sqrt{-9} \neq \sqrt{(-16)(-9)} = \sqrt{144} = 12$)

A **complex number** is any number in the form of $a + bi$, where a, b are real numbers, and i is the imaginary unit.

$\quad a + bi = c + di$ if and only if $a = c$ and $b = d$.

e.g. $\quad 5x - 2 = 3x + (2y + 4)i$

$\begin{array}{l|l} 5x - 2 = 3x & 0 = 2y + 4 \\ x = 1 & y = -2 \end{array}$

We can locate a complex number $a + bi$ in the complex plane the same way as we locate a point (a, b) in the rectangular plane.

A complex number $a + bi$ can also be represented as a vector $\overrightarrow{OA}$ whose initial point is the origin O and the end point is the point $A(a, b)$.

The **modulus** of a complex number is the magnitude (length) of the vector:

$$|a + bi| = \sqrt{a^2 + b^2}$$

e.g.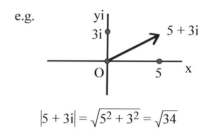

$|5 + 3i| = \sqrt{5^2 + 3^2} = \sqrt{34}$

20.2 Addition of Complex Numbers

(1) Algebraic Method

$$(a + bi) + (c + di) = (a + c) + (b + d)i$$

(2) Geometric Method

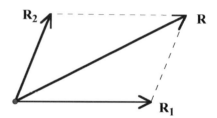

The sum **R** is the resultant vector.

e.g. $\quad (5 + 2i) + (-2 + 3i)$

represented algebraically:
$= (5 - 2) + (2i + 3i) = 3 + 5i$

represented graphically:

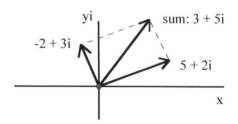

58.

III. Coordinate Geometry and Functions

20.3 Subtraction of Complex Numbers

Subtraction is the addition of an additive inverse (the opposite number).

(1) Algebraic Method

$$(a + bi) - (c + di) = (a - c) + (b - d)i$$

(2) Geometric Method

$\mathbf{R_1} - \mathbf{R_2} = \mathbf{R_1} + (-\mathbf{R_2}) = \mathbf{R}$
$-\mathbf{R_2}$ is a 180° rotation of $\mathbf{R_2}$ about the origin.

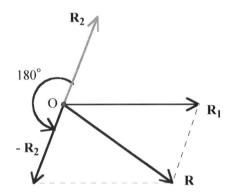

20.4 Multiplication of Complex Numbers

(1) Algebraic Method
Use the distributive property

$$(a + bi)(c + di) = (ac - bd) + (ad + bc)i$$

(2) Geometric Method
A complex number $\vec{OA}$ multiplied by a real number c is equivalent to a dilation of c.

$$c\,\vec{OA} \quad \underline{\;D\,c\;} \quad \vec{OA'}$$

A complex number $\vec{OA}$ multiplied by i is equivalent to a 90° rotation about the origin.

$$i\,\vec{OA} \quad \underline{\;R90°\;} \quad \vec{OA'}$$

e.g $(2 + i)(1 - 2i)$

(1) Algebraic Method
$$(2 + i)(1 - 2i)$$
$$= 2 - 4i + i - 2i^2$$
$$= 2 - 3i + 2$$
$$= 4 - 3i$$

(2) Geometric Method
$$(2 + i)(1 - 2i)$$
$$= 2(1 - 2i) + i(1 - 2i)$$

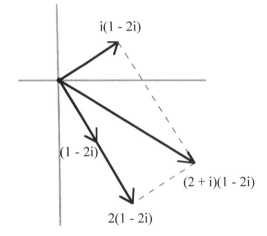

Conjugates:

$a + bi$ and $a - bi$ are conjugates.

The sum and the product of two conjugates are real numbers.

$$(a + bi) + (a - bi) = 2a$$
$$(a + bi)(a - bi) = a^2 - (bi)^2 = a^2 - b^2 i^2 = a^2 + b^2$$

Multiplicative Inverse:

e.g. Use the conjugate to write the multiplicative inverse of $2 + 4i$ in $a + bi$ form.

$$\frac{1}{2 + 4i} = \frac{1}{2 + 4i} \cdot \frac{2 - 4i}{2 - 4i} = \frac{2 - 4i}{4 + 16}$$
$$= \frac{2 - 4i}{20} = \frac{1}{10} - \frac{1}{5}i$$

e.g. The multiplicative inverse of $5i$ is

$$\frac{1}{5i} = \frac{1}{5i} \cdot \frac{i}{i} = -\frac{i}{5}$$

III. Coordinate Geometry and Functions

20.5 Division of Complex Numbers

Division is the multiplication of the reciprocal (the multiplicative inverse).

e.g. $\dfrac{8+i}{2-i}$

$= \dfrac{(8+i)(2+i)}{(2-i)(2+i)}$ $2+i$ is the conjugate of $2-i$

$= \dfrac{16 + 8i + 2i + i^2}{4 - i^2}$

$= \dfrac{15 + 10i}{5} = 3 + 2i$

20.6 The Midpoint of Two Complex Numbers

$A = x_1 + y_1 i, \quad B = x_2 + y_2 i$

$\dfrac{A+B}{2} = \dfrac{x_1 + x_2}{2} + \dfrac{y_1 + y_2}{2} i$

20.7 The Distance Between Two Complex Numbers

$A = x_1 + y_1 i, \quad B = x_2 + y_2 i$

$|A - B| = \sqrt{(x_2 - x_1)^2 + (y_2 - y_1)^2}$

20.8 The Modulus of the Product of Two Complex Numbers

$|A \cdot B| = |A| \cdot |B|$

e.g. $A = 2 + i, \quad B = 1 - 2i$

$|A \cdot B| = |(2+i)(1-2i)| = |4 - 3i|$
$= \sqrt{a^2 + b^2} = \sqrt{4^2 + (-3)^2} = 5$

$|A| \cdot |B| = |2 + i| \cdot |1 - 2i|$
$= \sqrt{2^2 + 1^2} \cdot \sqrt{1^2 + (-2)^2}$
$= \sqrt{5} \cdot \sqrt{5} = 5$

20.9 The Polar Form of a Complex Number

$Z = x + yi = r(\cos\theta + i\sin\theta)$

where r is the modulus that equals $|Z|$ and θ is the angle between the polar axis and the line $\overline{OP}$.

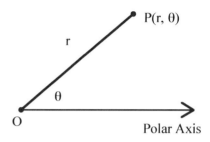

$x = r\cos\theta, \quad y = r\sin\theta$
$r = \sqrt{x^2 + y^2}$
$\tan\theta = \dfrac{y}{x}$ or $\theta = \tan^{-1}\left(\dfrac{y}{x}\right)$

e.g.
Convert the complex number to rectangular form:

$Z = 5(\cos 135° + i \sin 135°)$
$= 5\left(-\dfrac{\sqrt{2}}{2} + \dfrac{\sqrt{2}}{2} i\right)$
$= -\dfrac{5\sqrt{2}}{2} + \dfrac{5\sqrt{2}}{2} i$

e.g.
Convert the complex number to polar form:

$Z = \sqrt{3} - i$
$r = \sqrt{x^2 + y^2} = \sqrt{3 + 1} = 2$
$\tan\theta = \dfrac{-1}{\sqrt{3}} = -\dfrac{\sqrt{3}}{3}$
Reference angle $\theta_r = 30°$
Since Z is in the 4th quadrant,
$\theta = 360° - 30° = 330°$

$Z = 2(\cos 330° + i \sin 330°)$

III. Coordinate Geometry and Functions

20.10 Multiplication and Division in Polar Form

Let $Z_1 = r_1(\cos\theta_1 + i\sin\theta_1)$
$Z_2 = r_2(\cos\theta_2 + i\sin\theta_2)$, then

$$Z_1 \cdot Z_2 = r_1 \cdot r_2[\cos(\theta_1 + \theta_2) + i\sin(\theta_1 + \theta_2)]$$

$$\frac{Z_1}{Z_2} = \frac{r_1}{r_2}[\cos(\theta_1 - \theta_2) + i\sin(\theta_1 - \theta_2)]$$

where $Z_2 \neq 0$.

20.11 Powers of a Complex Number in Polar Form

Let $Z = r(\cos\theta + i\sin\theta)$, then

$$Z^n = r^n(\cos n\theta + i\sin n\theta)$$

where n is a positive integer.

20.12 Roots of a Complex Number in Polar Form

Let $Z = r(\cos\theta + i\sin\theta)$,
then the nth roots of Z are:

$$z_k = \sqrt[n]{r}\left(\cos\frac{\theta + 2k\pi}{n} + i\sin\frac{\theta + 2k\pi}{n}\right)$$

or $z_k = \sqrt[n]{r}\left(\cos\frac{\theta + k \cdot 360°}{n} + i\sin\frac{\theta + k \cdot 360°}{n}\right)$

where $k = 0, 1, 2, \cdots, n-1$.

e.g. Find the complex cube roots of
$Z = 8(\cos 60° + i\sin 60°)$

$z_0 = \sqrt[3]{8}\left(\cos\frac{60°}{3} + i\sin\frac{60°}{3}\right)$
$= 2(\cos 20° + i\sin 20°)$

$z_1 = \sqrt[3]{8}\left(\cos\frac{60° + 360°}{3} + i\sin\frac{60° + 360°}{3}\right)$
$= 2(\cos 140° + i\sin 140°)$

$z_2 = \sqrt[3]{8}\left(\cos\frac{60° + 720°}{3} + i\sin\frac{60° + 720°}{3}\right)$
$= 2(\cos 260° + i\sin 260°)$

20.13 Revisit Quadratic Equations

$ax^2 + bx + c = 0$ where $a \neq 0$

The sum of the roots $x_1 + x_2 = -\dfrac{b}{a}$

The product of the roots $x_1 \cdot x_2 = \dfrac{c}{a}$

$$x = \frac{-b \pm \sqrt{b^2 - 4ac}}{2a}$$

where $b^2 - 4ac$ is called discriminant Δ

$\Delta > 0$ two unequal real roots (two x intercepts)
$\Delta = 0$ two equal real roots (one x intercept)
$\Delta < 0$ no real roots (no x intercept)
when $\Delta < 0$, it has two imaginary roots in the
form of conjugates: $x_1 = a + bi$ and $x_2 = a - bi$

e.g. $x^2 + 2x + 2 = 0$ $x_1 = -1 + i$, $x_2 = -1 - i$

e.g. $2x^2 + 5x + 25 = 0$

$x = \dfrac{-b \pm \sqrt{b^2 - 4ac}}{2a}$

$= \dfrac{-5 \pm \sqrt{5^2 - 4 \cdot 2 \cdot 25}}{2 \cdot 2}$

$= \dfrac{-5 \pm \sqrt{5^2 - 4 \cdot 2 \cdot 25}}{2 \cdot 2}$

$= \dfrac{-5 \pm \sqrt{-175}}{4}$

$= -\dfrac{5}{4} \pm \dfrac{5\sqrt{7}}{4}i$

Factoring extended to complex numbers:

$$x^2 + a^2 = (x + ai)(x - ai)$$

e.g. $x^2 + 25 = 0$
$(x + 5i)(x - 5i) = 0$
$x = -5i$, $x = 5i$

III. Coordinate Geometry and Functions

20.14 Fundamental Theorem of Algebra

A polynomial equation of degree n has n complex roots which are real, imaginary, or both. These zeros may be repeated.

The real roots are the x-intercepts and the imaginary roots are paired complex conjugates when all coefficients are real.

An odd degree polynomial function with real coefficients has at least one real zero.

20.15 Linear Factorization Theorem

If f(x) is a polynomial of degree of n, then

$$f(x) = a(x - z_1)(x - z_2) \cdots (x - z_n)$$

where a is the leading coefficient of f(x) and z_1, z_2, $\cdots$, z_n are complex numbers that are zeros of f(x). The z_i is not necessarily distinct.

e.g $x^3 - 3x^2 + 4x - 12 = 0$
$x^2(x - 3) + 4(x - 3) = 0$
$(x - 3)(x^2 + 4) = 0$

$x - 3 = 0$	$x^2 + 4 = 0$
$x = 3$	$x^2 = -4$
	$x = \pm\sqrt{-4}$
	$x = \pm 2i$

Solution: { 3, -2i, 2i }
$f(x) = (x - 3)(x + 2i)(x - 2i)$

e.g. $x^4 + 3x^2 = 4$
$x^4 + 3x^2 - 4 = 0$
$(x^2 + 4)(x^2 - 1) = 0$

$x^2 + 4 = 0$	$x^2 - 1 = 0$
$x = \pm 2i$	$(x + 1)(x - 1) = 0$
	$x = -1, x = 1$

Solution set: { -1, 1, -2i, 2i }
$f(x) = (x + 1)(x - 1)(x + 2i)(x - 2i)$

e.g. Solve $x^3 - 1 = 0$ in complex number domain.

Method 1:
$(x - 1)(x^2 + x + 1) = 0$
$x - 1 = 0$
$x = 1$
$x^2 + x + 1 = 0$
$x = \dfrac{-1 \pm \sqrt{1 - 4}}{2}$
$= \dfrac{-1 \pm i\sqrt{3}}{2}$

Method 2: Use polar form to solve the equation.

$x^3 = 1$
$= 1 + 0 \cdot i$
$r = 1, \theta = 0°, n = 3, k = 0, 1, 2$

$x_k = \sqrt[n]{r} \left(\cos\dfrac{\theta + k \cdot 360°}{n} + i \sin\dfrac{\theta + k \cdot 360°}{n}\right)$

$x_0 = \sqrt[3]{1} \left(\cos\dfrac{0° + 0 \cdot 360°}{3} + i \sin\dfrac{0° + 0 \cdot 360°}{3}\right)$
$= \cos 0° + i \sin 0° = 1$

$x_1 = \sqrt[3]{1} \left(\cos\dfrac{0° + 360°}{3} + i \sin\dfrac{0° + 360°}{3}\right)$
$= \cos 120° + i \sin 120°$
$= -\dfrac{1}{2} + \dfrac{\sqrt{3}}{2} i$

$x_2 = \sqrt[3]{1} \left(\cos\dfrac{0° + 2 \cdot 360°}{3} + i \sin\dfrac{0° + 2 \cdot 360°}{3}\right)$
$= \cos 240° + i \sin 240°$
$= -\dfrac{1}{2} - \dfrac{\sqrt{3}}{2} i$

Both have the same solution set:
$\{1, -\dfrac{1}{2} - \dfrac{\sqrt{3}}{2} i, -\dfrac{1}{2} + \dfrac{\sqrt{3}}{2} i\}$

III. Coordinate Geometry and Functions

21. GRAPHIC SOLUTIONS OF LOCI AND SYSTEM OF EQUATIONS

21.1 Equations of Loci

(1) The center-radius equation of a circle with radius r and center (h, k)

$$(x - h)^2 + (y - k)^2 = r^2$$

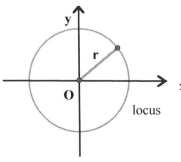

$$x^2 + y^2 = r^2$$

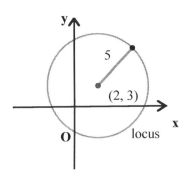

$$(x - 2)^2 + (y - 3)^2 = 5^2$$

(2) The equation of the perpendicular bisector

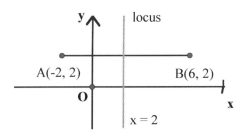

$x = 2$ is the equation of the perpendicular bisector of $\overline{AB}$.

Find the equation of the locus of points equidistant from points A(-2, 2) and B(4, -2).

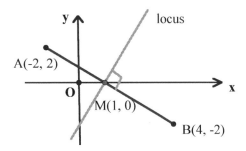

Find the midpoint of $\overline{AB}$:
$$M(\bar{x}, \bar{y}) = M(\frac{-2 + 4}{2}, \frac{2 + (-2)}{2}) = M(1, 0)$$

Find the slope of $\overline{AB}$:
$$m_1 = \frac{-2 - 2}{4 - (-2)} = \frac{-4}{6} = -\frac{2}{3}$$

the slope of the perpendicular line:
$$m_2 = -\frac{1}{m_1} = \frac{3}{2}$$

the equation of the perpendicular bisector:
the slope is $\frac{3}{2}$ and passing through midpoint (1, 0)

the point-slope form: $y - y_1 = m(x - x_1)$
$$y - 0 = \frac{3}{2}(x - 1), \text{ or}$$

slope-intercept form: $y = \frac{3}{2}x - \frac{3}{2}$

(3) The equation of the angle bisector

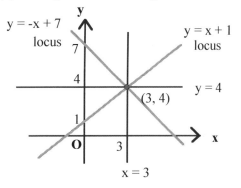

$y = x + 1$ and $y = -x + 7$ are the loci of the points equidistant from lines $x = 3$ and $y = 4$.
Hint: $m = \pm 1$ and passing through point (3, 4)

III. Coordinate Geometry and Functions

21.2 Graphic Solutions of System of Equations

(1) Linear System:

e.g. $\quad x + y = 7$
$\quad\quad 2x - y = 2$

rewrite in the slope-intercept form:
$y = -x + 7$
$y = 2x - 2$

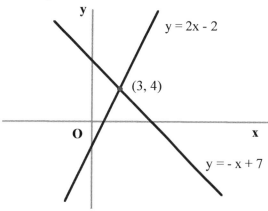

The solution of the system of equations is the intersection point (3, 4) of the two lines.

(2) Quadratic-Linear System:

e.g. $\quad y = x^2 - 8 \quad$ (1)
$\quad\quad y + 5 = 2x \quad$ (2)

rewrite Eq. (2) in the slope-intercept form:
$y = 2x - 5$

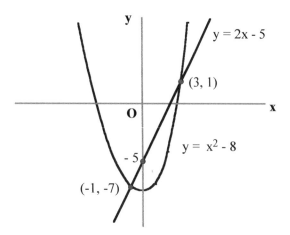

The solution of the system of equations are the intersection points (3, 1) and (-1, -7).

21.3 Graphic Solutions of System of Inequalities

e.g. The solution of the system
$\quad x > 2 \quad\quad$ (1)
$\quad 2x - y \geq 6 \quad$ (2)

rewrite (2) in the slope-intercept form
$-y \geq -2x + 6$
$y \leq 2x - 6 \quad$ inequality sign reversed

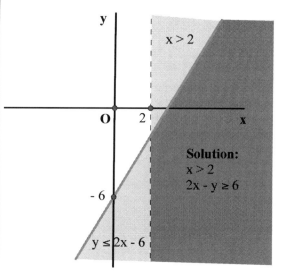

Solution:
$x > 2$
$2x - y \geq 6$

e.g. The solution of the system
$\quad y \geq x^2 - 8 \quad$ (1)
$\quad y + 5 \leq 2x \quad$ (2)

rewrite inequality (2) in the slope-intercept form: $\quad y \leq 2x - 5$

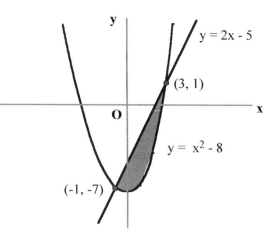

The solution of the system of inequalities is the shaded region including the border.

IV. Trigonometry

22. TRIGONOMETRIC FUNCTIONS

22.1 Degrees and Radians

The degree measure of a circle is 360.
The degree measure of a semicircle is 180.
Degree °, Minute ', Second "
$1° = 60'$, $1' = 60''$

e.g. $0.3° = 0.3 \cdot 60 = 18'$
$45' = \dfrac{45}{60} = 0.75°$

Radian is a different unit to measure the angle and the arc.
The radian measure of a circle is 2π.
The radian measure of a semicircle is π.
2π radians = $360°$, π radians = $180°$
1 radian = $\dfrac{180°}{\pi} \approx 57.3°$

We usually omit the word "radian".
$2\pi = 360°$, $\pi = 180°$

Conversion:
use $\dfrac{\pi}{180°} = 1$ or $\dfrac{180°}{\pi} = 1$

e.g. $60° = 60° \cdot \dfrac{\pi}{180°} = \dfrac{\pi}{3}$

e.g. $\dfrac{\pi}{4} = \dfrac{\pi}{4} \cdot \dfrac{180°}{\pi} = 45°$

22.2 Coterminal Angles

Coterminal angles have the same terminal side. They differ 360° or a multiple of 360°.

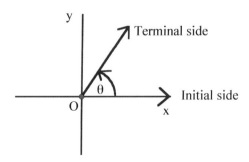

Counterclockwise angles are positive;
Clockwise angles are negative.

e.g. 30°, 390°, and 750° are coterminal angles.

e.g. -60° and 300° are coterminal angles.

22.3 Arc Length

$s = r \cdot \theta$ where r is radius and θ is in radians

e.g.
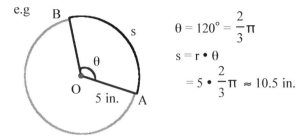

$\theta = 120° = \dfrac{2}{3}\pi$
$s = r \cdot \theta$
$= 5 \cdot \dfrac{2}{3}\pi \approx 10.5$ in.

These problems can also be solved by using fractions or ratios of the circle.

22.4 Trigonometric Fuctions

Trigonometric Ratios and Basic Functions

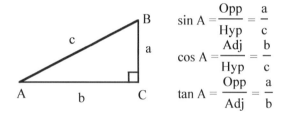

$\sin A = \dfrac{Opp}{Hyp} = \dfrac{a}{c}$

$\cos A = \dfrac{Adj}{Hyp} = \dfrac{b}{c}$

$\tan A = \dfrac{Opp}{Adj} = \dfrac{a}{b}$

Reciprocal Functions:

$\cot A = \dfrac{1}{\tan A}$, $\sec A = \dfrac{1}{\cos A}$, $\csc A = \dfrac{1}{\sin A}$

Exact Values to Remember:

θ (degree)	0°	30°	45°	60°	90°
θ	0	π/6	π/4	π/3	π/2
sinθ	0	1/2	$\sqrt{2}/2$	$\sqrt{3}/2$	1
cosθ	1	$\sqrt{3}/2$	$\sqrt{2}/2$	1/2	0
tanθ	0	$\sqrt{3}/3$	1	$\sqrt{3}$	undefined

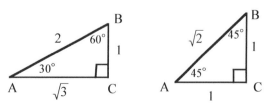

IV. Trigonometry

22.5 Pythagorean Triples

3, 4, 5 and 5, 12, 13

e.g. If $\sin \theta = \frac{3}{5}$, then $\cos \theta = \frac{4}{5}$, $\tan \theta = \frac{3}{4}$

If $\tan \theta = \frac{5}{12}$, then $\sin \theta = \frac{5}{13}$, $\cos \theta = \frac{12}{13}$

22.6 Cofunctions

$\sin \theta$ and $\cos \theta$ are cofunctions.
$\tan \theta$ and $\cot \theta$ are cofunctions.
$\sec \theta$ and $\csc \theta$ are cofunctions.

Any trigonometric function of an acute angle is equal to the cofunction of its complement.

$\sin \theta = \cos (90° - \theta)$
$\tan \theta = \cot (90° - \theta)$
$\sec \theta = \csc (90° - \theta)$

e.g. $\sin 30° = \cos 60°$

e.g. If $\sin A = \cos B$, then $A + B = 90°$

e.g. If $\sin 2A = \cos 4A$, find A
    $2A + 4A = 90°$ $A = 15°$

22.7 Unit Circle

Center at the origin ; Radius of one unit.

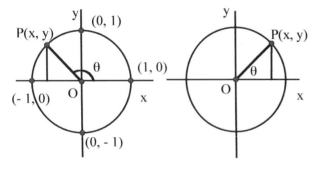

For Angles of Any Degree:

$\cos \theta = x$, $\sin \theta = y$, $\tan \theta = \frac{y}{x}$

Use unit circle to find the values of trigonometric functions at **quadrantal angles** (0°, 90°, 180°, 270°, 360°, etc.).

e.g. $\cos 180° = -1$, $\sin 180° = 0$, $\tan 180° = 0$
     $\cos 270° = 0$, $\sin 270° = -1$, $\tan 270° =$ undefined

Use line segments to represent the trigonometric functions in the first quadrant of the unit circle.

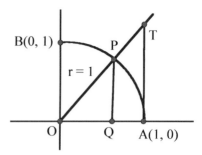

$PQ = \sin \theta$, $OQ = \cos \theta$, $TA = \tan \theta$

In general, a point is not on a unit circle:

$\cos \theta = \dfrac{x}{\sqrt{x^2 + y^2}}$, $\sin \theta = \dfrac{y}{\sqrt{x^2 + y^2}}$, $\tan \theta = \dfrac{y}{x}$

Coterminal angles have the same trigonometric function values.

e.g. $\sin 390° = \sin 30° = \dfrac{1}{2}$

$\cos 405° = \cos 45° = \dfrac{\sqrt{2}}{2}$

22.8 The Signs of Trigonometric Functions

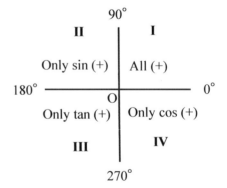

IV. Trigonometry

22.9 Reference Angle

The **reference angle** for any angle in standard position is an acute angle formed by the terminal side of the given angle and the x-axis.

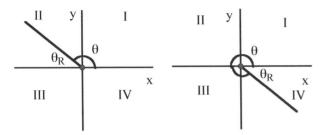

We use the following table to find the reference angle θ_R for a given angle θ, then use θ_R to find the values of the trigonometric functions.

Quadrant	I	II	III	IV
$\theta_R =$	θ	$180° - \theta$	$\theta - 180°$	$360° - \theta$
$\sin \theta =$	$\sin \theta_R$	$\sin \theta_R$	$- \sin \theta_R$	$- \sin \theta_R$
$\cos \theta =$	$\cos \theta_R$	$- \cos \theta_R$	$- \cos \theta_R$	$\cos \theta_R$
$\tan \theta =$	$\tan \theta_R$	$- \tan \theta_R$	$\tan \theta_R$	$- \tan \theta_R$

e.g.

$\cos 120° = - \cos 60° = -\dfrac{1}{2}$ ($\theta_R = 180° - 120° = 60°$)

$\cos 240° = - \cos 60° = -\dfrac{1}{2}$ ($\theta_R = 240° - 180° = 60°$)

$\cos 300° = \cos 60° = \dfrac{1}{2}$ ($\theta_R = 360° - 300° = 60°$)

e.g.
If $\sin \theta = \dfrac{5}{13}$ and θ is in the quadrant II, find the values of $\cos \theta$ and $\tan \theta$.

First find all the values of the functions about the θ_R.

$\sin \theta_R = \dfrac{5}{13}$, $\cos \theta_R = \dfrac{12}{13}$, $\tan \theta_R = \dfrac{5}{12}$

$\cos \theta$ and $\tan \theta$ are both negative in quadrant II.

$\cos \theta = -\dfrac{12}{13}$, $\tan \theta = -\dfrac{5}{12}$

22.10 Basic Applications

Determine the right triangle and use the trigonometric ratios to solve the problem.

e.g. Find the height of a tree.

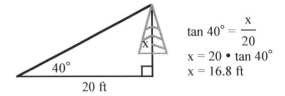

$\tan 40° = \dfrac{x}{20}$

$x = 20 \cdot \tan 40°$

$x = 16.8$ ft

e.g. A ladder is leaning against a vertical wall, making an angle of 60° with the ground and reaching a height of 12 feet on the wall.

Find, to the nearest foot, the length of the ladder.
Find, to the nearest foot, the distance from the base of the ladder to the wall.

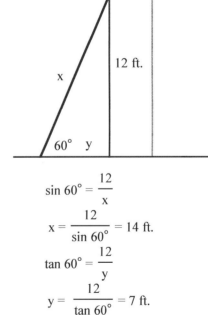

$\sin 60° = \dfrac{12}{x}$

$x = \dfrac{12}{\sin 60°} = 14$ ft.

$\tan 60° = \dfrac{12}{y}$

$y = \dfrac{12}{\tan 60°} = 7$ ft.

e.g.

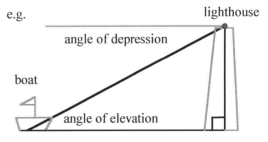

angle of elevation = angle of depression

67.

IV. Trigonometry

23. TRIGONOMETRIC GRAPHS

23.1 Graphs of Trigonometric Functions

$y = \sin x$

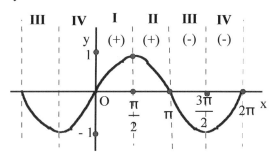

Domain: { x | x all real numbers }
Range: { y | -1 ≤ y ≤ 1 }

$y = \cos x$

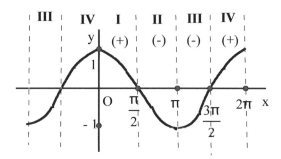

Domain: { x | x all real numbers }
Range: { y | -1 ≤ y ≤ 1 }

$y = \tan x$

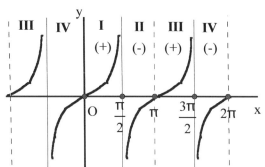

Domain: { x | x ≠ $\frac{\pi}{2}$ + nπ for n an integer }
Range: { y | y all real numbers }

23.2 Graphs of the Reciprocal Functions

when $f(x) \to 0$, $\frac{1}{f(x)} \to \infty$

$0 < f(x) < 1$, $\frac{1}{f(x)} > 1$

$f(x) = 1$, $\frac{1}{f(x)} = 1$

$f(x) > 1$, $0 < \frac{1}{f(x)} < 1$

$f(x) \to \infty$, $\frac{1}{f(x)} \to 0$, etc.

e.g. $y = \csc x = \frac{1}{\sin x}$

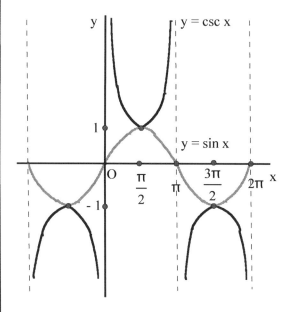

Domain: { x | x : all real numbers except nπ }
Range: { y | y ≤ -1 or y ≥ 1 }

Use graphing calculator to see y = cot x and y = sec x.

IV. Trigonometry

23.3 Amplitude, Period, and Frequency

(1) y = asin bx   and   y = acos bx

For $|a|, |b| > 1$, the graph of $y = a\sin bx$ is the graph of $y = \sin x$ stretched in vertical direction by a factor $|a|$ and shrunk in horizontal direction by a factor of $\left|\dfrac{1}{b}\right|$.

For $|a|, |b| < 1$, the graph of $y = a\sin bx$ is the graph of $y = \sin x$ shrunk in vertical direction by a factor $|a|$ and stretched in horizontal direction by a factor of $\left|\dfrac{1}{b}\right|$.

Amplitude: $A = \dfrac{y_{max} - y_{min}}{2} = |a|$,

Frequency: $F = |b|$

Period: $P = \dfrac{2\pi}{F} = \dfrac{2\pi}{|b|}$

e.g. $y = \cos x$
   amplitude = 1, frequency = 1, period = 2π

e.g. $y = -3\sin 2x$
   amplitude = 3, frequency = 2, period = $\dfrac{2\pi}{2} = \pi$

(2) y = tan bx

Frequency: $F = |b|$,   Period: $P = \dfrac{\pi}{F} = \dfrac{\pi}{|b|}$

e.g. $y = \tan x$ period = π,
   therefore $\tan x = \tan(x + \pi)$

e.g. $y = \tan 2x$ period = $\dfrac{\pi}{2}$
   therefore $\tan 2x = \tan 2(x + \dfrac{\pi}{2}) = \tan(2x + \pi)$

23.4 The Graph of y = asin b(x - c) + d

The graph of $y = a\sin b(x - c) + d$ is the graph of $y = a\sin bx$ shifted c units to the right and d units up. c is called the **phase shift** and $y = d$ is called the **midline**.

$$d = \dfrac{y_{max} + y_{min}}{2}$$

e.g. Graph $y = -2\cos(\dfrac{1}{2}x) + 1$

Amplitude: $A = |a| = 2$,

Frequency: $F = |b| = \dfrac{1}{2}$,

Period $P = \dfrac{2\pi}{F} = 4\pi$

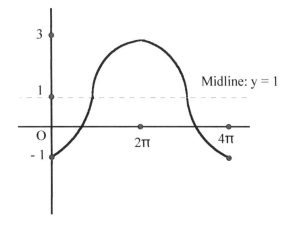

Midline: $y = 1$

e.g. Graph $y = 5\sin 2(x + \dfrac{\pi}{4}) + 3$

Amplitude = 5, Frequency = 2, Period = π
Midline: $y = 3$

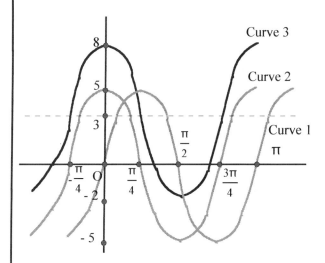

Curve 1: $y = 5\sin 2x$,

Curve 2: $y = 5\sin 2(x + \dfrac{\pi}{4})$, shift $\dfrac{\pi}{4}$ to the left

   (phase shift $-\dfrac{\pi}{4}$)

Curve 3: $y = 5\sin 2(x + \dfrac{\pi}{4}) + 3$, shift 3 units up

IV. Trigonometry

23.5 Inverse Trigonometric Functions

$y = \arcsin x$ or $y = \sin^{-1} x$

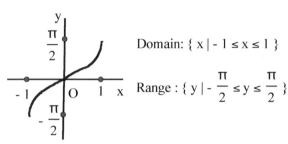

Domain: $\{ x \mid -1 \leq x \leq 1 \}$

Range: $\{ y \mid -\dfrac{\pi}{2} \leq y \leq \dfrac{\pi}{2} \}$

$y = \arccos x$ or $y = \cos^{-1} x$

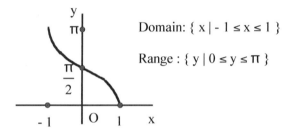

Domain: $\{ x \mid -1 \leq x \leq 1 \}$

Range: $\{ y \mid 0 \leq y \leq \pi \}$

$y = \arctan x$ or $y = \tan^{-1} x$

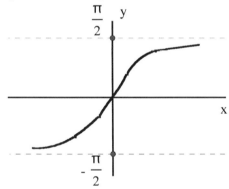

Domain: $\{ x \mid x$ all real numbers $\}$
Range: $\{ y \mid -\dfrac{\pi}{2} < y < \dfrac{\pi}{2} \}$

e.g. Find $\sin(\arccos \dfrac{1}{2})$

Let $\theta = \arccos \dfrac{1}{2}$ $(0 \leq \theta \leq \pi)$

$\cos \theta = \dfrac{1}{2}$

since $0 \leq \theta \leq \pi$, $\theta = 60°$

$\sin \theta = \dfrac{\sqrt{3}}{2}$

24. TRIGONOMETRIC IDENTITIES

$\tan \theta = \dfrac{\sin \theta}{\cos \theta}$, $\cot \theta = \dfrac{\cos \theta}{\sin \theta}$

$\sin^2 \theta + \cos^2 \theta = 1$
$\tan^2 \theta + 1 = \sec^2 \theta$
$\cot^2 \theta + 1 = \csc^2 \theta$

e.g. $\sin \theta = \dfrac{5}{13}$ and θ is in Quadrant II.
Find the value of (1) $\cos \theta$ and (2) $\tan \theta$.

(1) $(\dfrac{5}{13})^2 + \cos^2 \theta = 1$

$\cos \theta = -\dfrac{12}{13}$ ($\cos \theta$ is negative in Q II)

(2) $\tan \theta = \dfrac{\sin \theta}{\cos \theta} = \dfrac{5/13}{-12/13} = -\dfrac{5}{12}$

24.1 Trigonometric Identity Proofs

Show that both sides of the equation can be written in the same form.

e.g. Prove: $\dfrac{\sin^2 \theta}{1 - \cos \theta} = 1 + \cos \theta$

Since $\sin^2 \theta + \cos^2 \theta = 1$, we have
$\sin^2 \theta = 1 - \cos^2 \theta$

$\dfrac{1 - \cos^2 \theta}{1 - \cos \theta} = 1 + \cos \theta$

$\dfrac{(1 + \cos \theta)(1 - \cos \theta)}{1 - \cos \theta} = 1 + \cos \theta$

$1 + \cos \theta = 1 + \cos \theta$

24.2 Sum and Difference Formulas

$\sin(A + B) = \sin A \cos B + \cos A \sin B$
$\sin(A - B) = \sin A \cos B - \cos A \sin B$
$\cos(A + B) = \cos A \cos B - \sin A \sin B$
$\cos(A - B) = \cos A \cos B + \sin A \sin B$

By using the identities above, it is easy to prove the following identities:

$\sin(-\theta) = -\sin \theta$, $\cos(-\theta) = \cos \theta$, $\tan(-\theta) = -\tan \theta$
$\sin(90° - \theta) = \cos \theta$, $\cos(90° - \theta) = \sin \theta$

IV. Trigonometry

e.g. Find the exact value of $\cos 105°$.

$\cos 105° = \cos(60° + 45°)$
$= \cos 60° \cos 45° - \sin 60° \sin 45°$
$= \dfrac{1}{2} \cdot \dfrac{\sqrt{2}}{2} - \dfrac{\sqrt{3}}{2} \cdot \dfrac{\sqrt{2}}{2}$
$= \dfrac{\sqrt{2} - \sqrt{6}}{4}$

24.3 Double Angle Formulas

$\sin 2A = 2\sin A \cos A$

$\cos 2A = \cos^2 A - \sin^2 A$
$\cos 2A = 2\cos^2 A - 1$
$\cos 2A = 1 - 2\sin^2 A$

$\tan 2A = \dfrac{2\tan A}{1 - \tan^2 A}$

e.g. $\sin x = \dfrac{4}{5}$, x is an acute angle.
Find the value of $\cos 2x$ and $\sin 2x$.

(1) $\cos 2x = 1 - 2\sin^2 x$
$= 1 - 2\left(\dfrac{4}{5}\right)^2 = -\dfrac{7}{25}$

(2) $\cos x = \sqrt{1 - \sin^2 x} = \dfrac{3}{5}$ (acute angle)

$\sin 2x = 2 \sin x \cos x$
$= 2\left(\dfrac{4}{5}\right)\left(\dfrac{3}{5}\right) = \dfrac{24}{25}$

or $\sin 2x = \sqrt{1 - \cos^2 2x} = \dfrac{24}{25}$

($\sin 2x$ is positive when $2x$ is in either quadrant I or II.)

24.4 Half Angle Formulas

$\sin \dfrac{1}{2}A = \pm\sqrt{\dfrac{1 - \cos A}{2}}$

$\cos \dfrac{1}{2}A = \pm\sqrt{\dfrac{1 + \cos A}{2}}$

$\tan \dfrac{1}{2}A = \pm\sqrt{\dfrac{1 - \cos A}{1 + \cos A}}$

25. TRIGONOMETRIC EQUATIONS

We use the following table to find θ in different quadrants in terms of θ_R:

Quadrant	I	II	III	IV
θ	θ_R	$180° - \theta_R$	$180° + \theta_R$	$360° - \theta_R$

e.g. Solve for θ, $0° \leq \theta < 360°$
$\tan \theta + \sqrt{3} = 0$

$\tan \theta = -\sqrt{3}$
θ is in quadrant II and IV.
$\tan \theta_R = \sqrt{3}$
$\theta_R = 60°$
$\theta = 180° - 60° = 120°$
$\theta = 360° - 60° = 300°$
$\{120°, 300°\}$

e.g. Solve for θ, $0° \leq \theta < 180°$
$2\cos^2 \theta - \sin \theta = 1$

$2(1 - \sin^2 \theta) - \sin \theta = 1$
$2 - 2\sin^2 \theta - \sin \theta = 1$
$2\sin^2 \theta + \sin \theta - 1 = 0$
$(2\sin \theta - 1)(\sin \theta + 1) = 0$

$2\sin \theta - 1 = 0$	$\sin \theta + 1 = 0$
$\sin \theta = \dfrac{1}{2}$	$\sin \theta = -1$
	$\theta = 270°$
$\theta = 30°$	(not in the domain)

θ in quadrant I is also the reference angle θ_R.
Another angle in quadrant II, $\theta = 180 - \theta_R = 150°$
$\{30°, 150°\}$

e.g. Solve for θ, $0° \leq \theta < 360°$
$\cos 2\theta - \sin \theta = 1$

$1 - 2\sin^2 \theta - \sin \theta = 1$
$\sin \theta(2\sin \theta + 1) = 0$

$\sin \theta = 0$	$2\sin \theta + 1 = 0$
$\theta = 0°$ or $\theta = 180°$	$\sin \theta = -0.5$ no θ in quadrant I
	Solve $\sin \theta_R = 0.5$, $\theta_R = 30°$
	$\theta = 180° + 30° = 210°$, or
	$\theta = 360° - 30° = 330°$

$\{0°, 180°, 210°, 330°\}$

IV. Trigonometry

26. TRIGONOMETRIC APPLICATIONS

For any triangle:

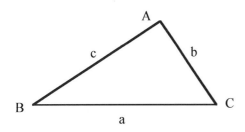

26.1 Area Formula

$$\text{Area} = \frac{1}{2}ab \sin C = \frac{1}{2}bc \sin A = \frac{1}{2}ca \sin B$$

26.2 Law of Sines

$$\frac{a}{\sin A} = \frac{b}{\sin B} = \frac{c}{\sin C}$$

26.3 Law of Cosines

$$c^2 = a^2 + b^2 - 2ab \cos C$$

$$\cos C = \frac{a^2 + b^2 - c^2}{2ab}$$

e.g. Find the area of $\triangle ABC$

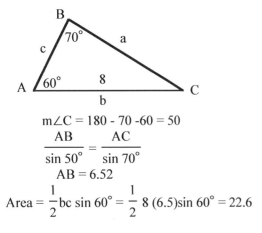

$$m\angle C = 180 - 70 - 60 = 50$$
$$\frac{AB}{\sin 50°} = \frac{AC}{\sin 70°}$$
$$AB = 6.52$$
$$\text{Area} = \frac{1}{2}bc \sin 60° = \frac{1}{2} \cdot 8 (6.5)\sin 60° = 22.6$$

e.g. Find x.

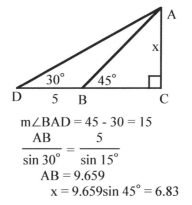

$$m\angle BAD = 45 - 30 = 15$$
$$\frac{AB}{\sin 30°} = \frac{5}{\sin 15°}$$
$$AB = 9.659$$
$$x = 9.659 \sin 45° = 6.83$$

e.g **Ambiguous Case:**

In $\triangle ABC$, $m\angle A = 30$, $AB = 12$, $BC = 7$, how many possible triangles can be constructed?

$$\frac{BC}{\sin A} = \frac{AB}{\sin C}$$
$$\frac{7}{\sin 30°} = \frac{12}{\sin C}$$
$$\sin C = \frac{6}{7}$$
$$m\angle C = 59 \text{ or } m\angle C = 180 - 59 = 121$$

check: $m\angle A + m\angle C = 30 + 59 < 180$ OK
check: $m\angle A + m\angle C = 30 + 121 < 180$ OK
Two possible triangles can be constructed.

e.g. Two forces of 10 lb. and 15 lb. act on a body. Their resultant is 20 lb. Find the angle between the two applied forces.

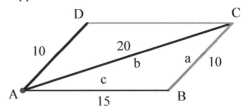

$BC = AD = 10$ (in a parallelogram opposite sides $\cong$)

In $\triangle ABC$, $\cos B = \dfrac{a^2 + c^2 - b^2}{2ac}$

$$= \frac{10^2 + 15^2 - 20^2}{2 \cdot 10 \cdot 15}$$

$$\cos B = -0.25$$
$$m\angle B = 104.5$$
$$m\angle BAD = 180 - 104.5 = 75.5$$

(Note: The angle between the two applied forces is $\angle BAD$, not $\angle B$.)

V. Probability and Statistics

27. PROBABILITY

27.1 Venn Diagram

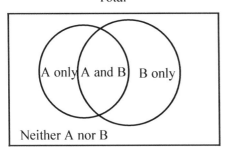

e.g. There are 22 vehicles. 8 of them are vans and 6 of them are red, 10 of them are neither vans nor red. How many red vans are there?

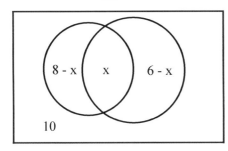

x: number of the red vans
$(8 - x) + x + (6 - x) + 10 = 22$
$-x + 24 = 22$
$x = 2$

27.2 Counting Principle (2 or more activities)

If the first activity can occur in M ways and the second activity can occur in N ways, then both activities can occur in M•N ways.

e.g. There are 3 doors to a building and 2 stairways to the second floor. We have 3•2 = 6 different ways to go to the second floor.

Tree Diagram:

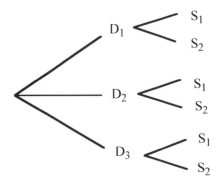

Sample Space: $\{(D_1, S_1), (D_1, S_2), (D_2, S_1), (D_2, S_2), (D_3, S_1), (D_3, S_2)\}$

27.3 Permutation and Combination

In a **permutation,** the order of the objects is important.

(1) The permutaion of n objects taken n at a time
$$_nP_n = n! = n(n-1)(n-2) \ldots 2 \cdot 1$$
n! is called **factorial**.

e.g. Five letters A, B, C, D, E have 5! different arrangements. ($5! = 5 \cdot 4 \cdot 3 \cdot 2 \cdot 1 = 120$)

(2) The permutation of n objects taken n at a time with r items identical : $\dfrac{n!}{r!}$

e.g. Five letters COLOR have $\dfrac{5!}{2!}$ different arrangements.

e.g. Seven letters FREEZER, three E's identical and two R's identical, have $\dfrac{7!}{3! \cdot 2!}$ different arrangements.

(3) The permutation of n objects taken r (r < n) at a time
$$_nP_r = n(n-1)(n-2) \ldots (n-r+1) \quad \text{(r factors)}$$

e.g. How many different arrangements of 1st, 2nd, and 3rd place are possible for 10 students?

$$_{10}P_3 = 10 \cdot 9 \cdot 8 = 720 \quad \text{(3 factors)}$$

V. Probability and Statistics

In a **combination,** the order of the objects does not matter.

e.g. (A, B, C) and (C, B, A) are considered same.

(4) The combination of n objects taken r at a time

$$_nC_r = \frac{_nP_r}{r!} \qquad (r \leq n)$$

$$_nC_n = 1, \quad _nC_0 = 1, \quad _nC_1 = n, \quad _nC_r = {}_nC_{n-r}$$

e.g. $\quad _{50}C_{48} = {}_{50}C_2 \quad$ (to simplify the calculation)

e.g. How many 3 player teams can be formed from 10 students?

$$_{10}C_3 = \frac{_{10}P_3}{3!} = \frac{10 \cdot 9 \cdot 8}{3 \cdot 2 \cdot 1} = 120$$

e.g. From 10 boys and 12 girls, how many different teams can be formed if 2 members must be boys and 3 members must be girls?

$$_{10}C_2 \cdot {}_{12}C_3 = \frac{10 \cdot 9}{2 \cdot 1} \cdot \frac{12 \cdot 11 \cdot 10}{3 \cdot 2 \cdot 1} = 9900$$

Using Graphing Calculator

e.g. $\quad 5! = 120$
[5] [MATH] PRB / 4: ! [ENTER]

e.g. $\quad _5P_3 = 60$
[5] [MATH] PRB / 2: $_nP_r$ [ENTER] 3 [ENTER]

e.g $\quad _7C_4 = 35$
[7] [MATH] PRB / 3: $_nC_r$ [ENTER] 4 [ENTER]

27.4 Probability

Sample Space: all possible outcomes
Event: the favorable outcomes

(1) The probability of a simple event:

$$P(E) = \frac{\text{number of outcomes of the event}}{\text{nubmer of outcomes of the sample space}}$$

$$P(E) = \frac{n(E)}{n(S)}$$

e.g. A bag contains 6 black balls and 4 white balls. What is the probability of selecting a black ball?

$$P(\text{Black}) = \frac{n(\text{Black})}{n(\text{Sample Space})} = \frac{6}{10}$$

Impossible Case: $\quad P(E) = 0$
Certain Case: $\quad P(E) = 1$
Negation: $\quad P(\text{Not } E) = 1 - P(E)$

e.g. If P(rain) = 30%,
then P(Not rain) = 1 - P(rain) = 70%

(2) The probability of a single event with two conditions A and B:

$$P(A \text{ and } B) = P(A) \cdot P(B)$$

e.g. 52 cards are shuffled and one chosen at random.

$$P(K \text{ and red}) = P(K) \cdot P(\text{red}) = \frac{4}{52} \cdot \frac{1}{2} = \frac{2}{52}$$

(3) The probability of a single event that satisfies condition A or condition B:

$$P(A \text{ or } B) = P(A) + P(B) - P(A \text{ and } B)$$

For disjoint sets A and B, we have P(A and B) = 0, then

$$P(A \text{ or } B) = P(A) + P(B)$$

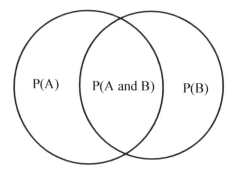

V. Probability and Statistics

e.g. 52 cards, P(K or red)
$$= P(K) + P(red) - P(K \text{ and red})$$
$$= \frac{4}{52} + \frac{26}{52} - \frac{2}{52}$$
$$= \frac{28}{52}$$

e.g. 52 cards, P(K or J)
$$= P(K) + P(J)$$
$$= \frac{4}{52} + \frac{4}{52}$$
$$= \frac{8}{52}$$
here P(K and J) = 0

(4) Counting Principle for Probability

When A and B are independent events, the compound event of A and B has the probability

$$P(A \text{ and } B) = P(A) \cdot P(B)$$

e.g. A fair die and a fair coin are tossed. What is the probability of obtaining a 2 on the die and a head on the coin?

$$P(2) = \frac{1}{6} \text{ and } P(head) = \frac{1}{2}$$
$$P(2 \text{ and head}) = P(2) \cdot P(head)$$
$$= \frac{1}{6} \cdot \frac{1}{2}$$
$$= \frac{1}{12}$$

e.g. There are 4 students. What is the probability of the tallest one in first place (A) and the shortest one in last place (B)?

$$P(A \text{ and } B) = P(A) \cdot P(B) = \frac{1}{4} \cdot \frac{1}{3} = \frac{1}{12}$$

27.5 Binomial Probability (Bernoulli Experiment)

If the probability of success is p and the probability of failure is q = 1 - p, then the probability of exactly r successes in n independent trials is

$$_nC_r \, p^r \, q^{n-r}$$

e.g. The probability of rain on any given day is 0.3.

(1) The probability of rain on exactly 2 of 7 days is
$$P(2) = {_7C_2}(0.3)^2(0.7)^5 \qquad (q = 1 - 0.3 = 0.7)$$

(2) The probability of rain at most 2 of 7 days is
$$P(\text{at most } 2) = P(0) + P(1) + P(2)$$
"at most 2 days" is same as "no more than 2 days"

(3) The probability of rain at least 4 of 7 days is
$$P(\text{at least } 4) = P(4) + P(5) + P(6) + P(7)$$
"at least 4 days" is same as "no less than 4 days"

(4) $P(0) + P(1) + P(2) + \cdots + P(6) + P(7) = 1$

e.g. $P(0) + P(1) + P(2) + \cdots + P(6) = 1 - P(7)$

27.6 Binomial Expansions

$$(x + y)^n = {_nC_0}x^n y^0 + {_nC_1}x^{n-1}y^1 + {_nC_2}x^{n-2}y^2 \cdots {_nC_n}x^0 y^n$$

There are n + 1 terms in the expansion.

The rth term is:

$$_nC_{(r-1)}\, x^{n-(r-1)}\, y^{(r-1)}$$

e.g. $(x+y)^4 = 1 \cdot x^4 + 4 \cdot x^3 \cdot y + 6 \cdot x^2 \cdot y^2 + 4 \cdot x \cdot y^3 + 1 \cdot y^4$

e.g. $(x+y)^8$
(1) the first term: (n = 8, r = 1, r - 1 = 0)
$$_8C_0 x^{8-0} y^0 = 1 \cdot x^8 \cdot 1 = x^8$$

(2) the last term: (n = 8, r = 9, r - 1 = 8)
$$_8C_8 x^{8-8} y^8 = 1 \cdot x^0 \cdot y^8 = y^8$$

(3) the middle term: (n = 8, r = 5, r - 1 = 4)
$$_8C_4 x^{8-4} y^4 = 70 x^4 y^4$$

e.g. $(2a - 1)^5$
The 3rd term is: (n = 5, r = 3, r - 1 = 2)
$$_5C_2 (2a)^{5-2}(-1)^2 = 10 \cdot (2a)^3 \cdot 1 = 80a^3$$

V. Probability and Statistics

28. STATISTICS (Univariate Data)

28.1 Common Methods of Collecting Data

Surveys: get information through questionnaires, interviews, etc.

Controlled Experiments: make cause and effect conclusions based on two groups of data, one of them serves as a benchmark

Observations: watch and study the phenomena, without influence on the responses

Size of the Data:

A **population** consists of the set of all items of interest. A **sample** is a subset of items chosen from a population. The sample must be large enough to be effective and must be chosen **randomly** to eliminate any **bias**.

We use sample data to estimate population characteristics.

28.2 Analyze Data

Mean, Median and Mode are the meaures of **central tendency of the data.**

Range, Quartile, Frequency and Standard Deviation are the measures of **dispersion of the data.**

First arrange the data in numerical order.

$$\text{Mean} = \text{Average} = \frac{\text{sum of the data values}}{\text{number of data items}}$$

Median: the middle value when data is arranged in order

Mode: the value that appears most often

Percentile: a number that tells what percent of the total number of data values are less than or equal to a given data point.

1st Quartile (25th percentile): the middle value of the lower half set of the data, aka. **Lower Quartile**

2nd Quartile (50th percentile): the **median**, aka. **Middle Quartile**

3rd Quartile (75th percentile): the middle value of the upper half set of the data, aka. **Upper Quartile**

Range: the difference between the highest value and the lowest value.

Interquartile Range (IQR): the difference between the 3rd quartile value and the 1st quartile value. It contains the middle 50% of the data values.

Outliers (Extreme Values): Any data value more than 1.5 IQR away from Q_1 or Q_3 is called an outlier.

Outliers can strongly affect the mean value. When outliers exist, use median to represent the central tendency of the data.

e.g. Analyze the grades:

$$78, 85, 81, 95, 61, 85, 75, 88, 72, 100$$

First rearrange the data in numerical order:
$$61, 72, 75, 78, 81, 85, 85, 88, 95, 100$$
(make sure the number of items are the same)

$$\text{Mean} = \frac{820}{10} = 82$$

$$\text{Median} = \frac{81 + 85}{2} = 83$$

(if the set has an even number of data values, take the average of the two middle values)

Mode = 85
Middle Quartile = Median = 83
Lower Quartile = 75
Upper Quartile = 88

Dot Plot:

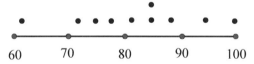

Box Plot (Five-number summary) :

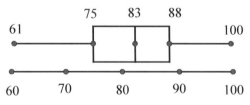

Range = 100 - 61 = 39
Interquartile Range = 88 - 75 = 13

V. Probability and Statistics

Frequency Table :

Interval	Frequency
61 - 70	1
71 - 80	3
81 - 90	4
91 - 100	2

Frequency Histogram :

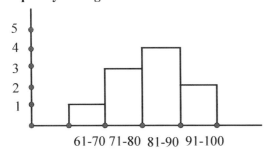

Relative Frequency Table :

Interval	Relative Frequency
61 - 70	1/10
71 - 80	3/10
81 - 90	4/10
91 - 100	2/10

Relative Frequency Histogram :

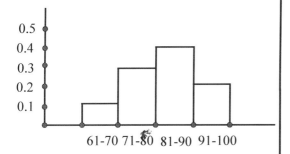

Cumulative Frequency Table :

Interval	Cumulative Frequency
61 - 70	1
61 - 80	4
61 - 90	8
61 - 100	10

Cumulative Frequency Histogram :

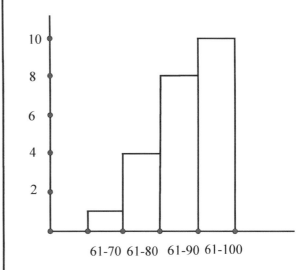

V. Probability and Statistics

28.3 Data Distribution

Although the five-number summary describes the distribution of the data, the most common method is the combination of the **mean** to measure the center and the **standard deviation** to measure the spread.

(1) Variance for Population:

$$v = \frac{1}{n} \sum_{i=1}^{n} (x_i - \bar{x})^2$$

(2) Standard Deviation for Population:

$$\delta = \sqrt{v} \qquad \text{(calculator symbol: } \delta_x\text{)}$$

n = number of data items,
$\bar{x}$ = mean, x_i = data values
Use graphing calculator to find $\bar{x}$ and δ_x.

Z-Score

$$\text{z-score} = \frac{x - \bar{x}}{\delta}$$

The z-score tells us how many standard deviations a data value x is above or below the mean.

e.g. Suppose the average score of the students is 80 with a standard deviation of 5. Find the z-scores of 90 and 65.

The z-score of 90 is $\frac{90 - 80}{5} = 2$

The z-score of 65 is $\frac{65 - 80}{5} = -3$

Alternatively, find the score with a z-score of -1.2.

$$\frac{x - 80}{5} = -1.2$$
$$x - 80 = -6$$
$$x = 74$$

Sample Survey

In the real world, we often use a sample from the population to study statistics.

(3) Variance for Sample:

$$v = \frac{1}{n-1} \sum_{i=1}^{n} (x_i - \bar{x})^2$$

(4) Standard Deviation for Sample:

$$S = \sqrt{v} \qquad \text{(calculator symbol: } S_x\text{)}$$

Use graphing calculator to find $\bar{x}$ and S_x.

Note the difference between δ_x and S_x in the formulas. With same set of data, S_x is slightly greater than δ_x since $\frac{1}{n-1}$ is greater $\frac{1}{n}$.

This difference is insignificant when the sample is large enough.

e.g. Find the mean and standard deviation of the following data set.

78, 85, 81, 95, 61, 85, 75, 88, 72, 100

Using Graphing Calculator

(1) Clear List L_1
[STAT] EDIT / 4: ClrList [ENTER] [2nd] [L_1]
[ENTER]

(2) Enter data to L_1
[STAT] EDIT / 1: Edit ... [ENTER]
    enter the above data into list L_1

(3) Display the One Variable Statistics
[STAT] CALC / 1: 1 - Var [ENTER] [2nd] [L_1]
[ENTER]

Mean: $\bar{x} = 82$
Standard Deviation: $\delta x = 10.74$
Standard Deviation for Samples: $S_x = 11.32$

V. Probability and Statistics

e.g. Find the mean and standard deviation for the following data set. x_i values, f_i frequency

x_i	92	87	82	77	72	67	62
f_i	2	3	6	9	10	6	4

(1) Clear List L_1 and List L_2
[STAT] EDIT / 4: ClrList [ENTER] [2nd]
[L_1] [,] [2nd] [L_2] [ENTER]

(2) Enter data to L_1 and L_2
[STAT] EDIT / 1: Edit ... [ENTER]
enter the x_i data into list L_1 and the f_i data into L_2

(3) Display the **One Variable** Statistics
[STAT] CALC / 1: 1 - Var [ENTER] [2nd]
[L_1] [,][2nd] [L_2] [ENTER]

$$\overline{x} = 75, \delta x = 7.89, S_x = 7.99,$$
$$Q_1 = 69.5, \text{Med} = 74.5, Q_3 = 82$$

(5) The Curve of the Data Distribution:

In most statistical studies, the number of data values is large. The relative frequency histogram becomes a smooth curve above the horizontal axis and the area underneath it is equal to 1.
This curve is called the **density curve**.

The median divides the the area under the curve in half.

The mean is the balance point in physics.

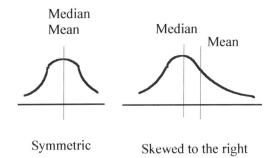

Symmetric   Skewed to the right

(6) Normal Distribution:

One particularly important class of density curves is the **normal curve** and its data distribution is called the **normal distribution**.

The properties of a normal curve:

(1) Symmetric.
(2) Mean, Median, and Mode have the same value.
(3) 68.2% of the data values between $\overline{x} - \delta$ and $\overline{x} + \delta$.
  95.4% of the data values between $\overline{x} - 2\delta$ and $\overline{x} + 2\delta$.
  99.8% of the data values between $\overline{x} - 3\delta$ and $\overline{x} + 3\delta$.

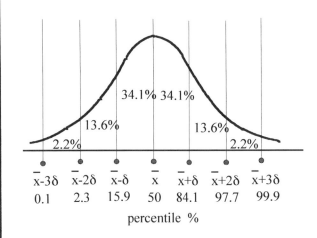

e.g. A normal distribution has a mean of 10.50 and a standard deviation of 0.75. What percent of the data values are in the range from 9.75 to 11.25 ?

$$9.75 = 10.50 - 0.75 = \overline{x} - \delta$$
$$11.25 = 10.50 + 0.75 = \overline{x} + \delta$$

There are 68.2% of the data values between 9.75 and 11.25.

e.g. The ages of the new teachers are normally distributed. 95.4% of the ages, centered about the mean, are between 24.6 and 37.4. Find the mean and the standard deviation.

$$\overline{x} = \frac{24.6 + 37.4}{2} = 31$$
$$(\overline{x} + 2\delta) - (\overline{x} - 2\delta) = 4\delta = 37.4 - 24.6 = 12.8$$
$$\delta = 3.2$$

V. Probability and Statistics

(7) Probability of Normal Distribution:

We can use z-score and the Standard Normal Probabilities table or use the graphing calculator to find the normal probabiity.

Three Types of Probabilities:
 P(less than a number)
 P(greater than a number) = 1 - P(less than a number)
 P(between two numbers A and B, A < B)
  = P(less than B) - P(less than A)

On TI 84 Graphing Calculator:
(1) Using the raw scores to find the probability:
[2nd] [DISTR] DISTR / 2: normalcdf(left bound, right bound, mean, standard deviation) [ENTER]
or Using the z-scores:
[2nd] [DISTR] DISTR / 2: normalcdf(left z bound, right z bound) [ENTER]

Note: When the values are relatively small, we can use - 999 for - ∞ and 999 for ∞

(2) Using given probability to find the raw scores:
[2nd] [DISTR] DISTR / 3: invNorm(area, mean, standard deviation) [ENTER]

e.g. The weight of a group of high school students is normally distributed with a mean of 152 lbs. and a standard deviation of 8 lbs.

(1) What is the probability of a student whose weight is less than 140 lbs.?

Method 1: z-score = $\dfrac{140 - 152}{8}$ = - 1.5

From the Standard Normal Probabilities table:
P(less than 140) = 0.0668

Method 2: Use TI 84 Graphing Calculator
 [2nd] [DISTR] DISTR / 2: normalcdf (-999, 140, 152, 8) [ENTER]
    0.0668
 or [2nd] [DISTR] DISTR / 2: normalcdf (-999, -1.5) [ENTER]
    0.0668

(2) What is the probability of a student whose weight is more than 162 lbs.?

Method 1: z-score = $\dfrac{162 - 152}{8}$ = 1.25

From the Standard Normal Probabilities table:
P(less than 162) = 0.8944
P(more than 162) = 1 - 0.8944 = 0.1056

Method 2: Use TI 84 Graphing Calculator
 [2nd] [DISTR] DISTR / 2: normalcdf (1.25, 999) [ENTER]
    0.1056

(3) What is the probability of a student whose weight is between 140 lbs and 162 lbs.?

Method 1: P(140 to 162)
 = P(less than 162) - P(less than 140)
 = 0.8944 - 0.0668
 = 0.8276

Method 2: Use TI 84 Graphing Calculator
 [2nd] [DISTR] DISTR / 2: normalcdf (-1.5, 1.25) [ENTER]
    0.8275

(4) The top 5% of the students will weigh more than what number of pounds?

Method 1: In the Standard Normal Probabilities table:
 Find the z-score for 0.95 = 1 - 0.05
  z-score = 1.645
  $\dfrac{x - 152}{8}$ = 1.645
  x = 165.16

Method 2: Use TI 84 Graphing Calculator
 [2nd] [DISTR] DISTR / 3: invNorm(0.95, 152, 8) [ENTER]
    165.16

V. Probability and Statistics

28.4 Statistical Inference

A summary measure calculated by using the data from a population is called a **population characteristic** or **parameter**.

A summary measure calculated by using the data from a sample is called a **sample statistic**.

We can use the data from the sample to estimate the population characteristics. Mean and proportion are two examples of population characteristics.

Sampling variability is the fact that the value of the sample statistic varies in repeated random sampling. The mean of the set of sample values is approximately equal to the population value.

When the sample size increases or the number of samples increases, the sampling variability decreases and the standard deviation of the sample values decreases.

e.g. Coin tossing problem
  Population proportion of Heads = 0.5

  Using the graphing calculator simulation:
  [MATH] / PRB / 7: randBin(trials, probabilty, simulations)[÷]trials [STO] [2nd] [L$_1$]

  (1). 20 tosses in each sample and repeat 40 times.
  [MATH] / PRB / 7: randBin(20, 0.5, 40) / 20 → L$_1$
  Calculate the mean and standard deviation of the sample proportions:
  [STAT] CALC / 1: 1 - Var [ENTER] [2nd] [L$_1$] [ENTER]
    mean $\bar{x}$ = 0.5125
    sample standard deviation S$_x$ = 0.1119

  (2). 50 tosses in each sample and repeat 40 times.
  [MATH] / PRB / 7: randBin(50, 0.5, 40) / 50 → L$_2$
  Calculate the mean and standard deviation of the sample proportions:
  [STAT] CALC / 1: 1 - Var [ENTER] [2nd] [L$_2$] [ENTER]
    mean $\bar{x}$ = 0.5075
    sample standard deviation S$_x$ − 0.0809
  Note: Each time the results may vary.

(1) Confidence Interval:

When we try to use statistics from one **simple random sample (SRS)** to estimate characteristics of the population, how much confidence do we have?

The confidence level can be expressed by the **confidence interval**:

  Estimate ± Margin of Error

The higher confidence interval requires larger margin of error.

When one standard deviation of the set of sample statistics is used for the margin of error, it is called **standard error,** which gives about a 68% confidence interval.

Two standard deviations is commonly used for the **margin of error**, which gives about a 95% confidence interval.

The standard deviation of the set of sample proportions:

$$\delta = \sqrt{\frac{p(1-p)}{n}}$$

  p: the population proportion (estimated by the sample proportion)
  n: the sample size

The standard deviation of the set of sample means:

$$\delta = \frac{s}{\sqrt{n}}$$

s: the standard deviation of the population (estimated by the standard deviation of the sample)
n: the sample size

Assumption of Sampling Distributions:

We assume that the sample be a simple random sample and be no more than 10% of the population. For sample proportion, np and n(1 - p) are both greater than 10. This means the sample size is large enough to have at least 10 successes and 10 failues. Then the distribution of the set of sample statistics can be considered normal.

V. Probability and Statistics

e.g. In a simple random sample of 50 students chosen, 28 students favor the new curriculum. Compute the margin of error and interpret the resulting interval in context.

Proportion: $p = \dfrac{28}{50} = 0.56$

We check: $np = 50 \cdot 0.56 = 28 > 10$
$n(1-p) = 50 \cdot 0.44 = 22 > 10$
$50 < 10\%$ of the population of students

The Estimated Margin of Error:

$ME = 2\sqrt{\dfrac{p(1-p)}{n}}$
$= 2\sqrt{\dfrac{0.56 \cdot 0.44}{50}}$
$= 0.14$

The resulting intervals:
0.56 ± 0.14 or
from 0.42 to 0.70

Since the lower end of the interval 0.42 is less than 0.50, we are not confident to say that more than half of the students favor the new curriculum.

e.g. In the above question, if 280 out of 500 students favor the new curriculum, what conclusion can we reach?

Proportion: $p = \dfrac{280}{500} = 0.56$

We check: $np = 500 \cdot 0.56 = 280 > 10$
$n(1-p) = 500 \cdot 0.44 = 220 > 10$
$500 < 10\%$ of the population of students

The Estimated Margin of Error:

$ME = 2\sqrt{\dfrac{p(1-p)}{n}}$
$= 2\sqrt{\dfrac{0.56 \cdot 0.44}{500}}$
$= 0.022$

The resulting intervals:
0.56 ± 0.022 or
from 0.538 to 0.582

Since the lower end of the interval 0.538 is higher than 0.50, we are confident to say that more than half of the students favor the new curriculum.

e.g. In a simple random sample of 50 LED bulbs tested for their life expectancy, the mean of the sample is 9800 hours with a standard deviation of 80 hours. Compute the margin of error and interpret the resulting interval in context.

We check: the sample size $50 < 10\%$ of the population.

The Estimated Margin of Error:

$ME = 2\dfrac{s}{\sqrt{n}}$
$= 2\dfrac{80}{\sqrt{50}}$
$= 22.6$

We are confident to say that the population mean of the life expectancy of the bulbs is

9800 ± 22.6 or
from 9777.4 to 9822.6

e.g. In the above question, if 500 LED bulbs are tested and we have the same sample mean and standard deviation, what conclusion can we reach?

We check: the sample size $500 < 10\%$ of the population.

The Estimated Margin of Error:

$ME = 2\dfrac{s}{\sqrt{n}}$
$= 2\dfrac{80}{\sqrt{500}}$
$= 7.2$

We are confident to say that the population mean of the life expectancy of the bulbs is

9800 ± 7.2 or
from 9792.8 to 9807.2

Note: When the sample size increases, the margin of error decreases.

V. Probability and Statistics

(2) Significance Level:

When we do an experiment, we observe the difference of the results from the treated group and the controlled group. How do we determine that the difference is due to the treatment, not due to the chance behavior?

A statement of "no effect or no difference between the two groups in the experiment" is called the **null hypothesis**, denoted as H_0.
A statement of "a difference between the two groups in the experiment" is called the **alternative hypothesis**, denoted as H_a.

The **significance level** is used as evidence to reject the null hypothesis H_0. The significance level is the probabilty of getting these results. When the significance level is about 5% or less, it is considered as an extreme case caused by the treatment, not by chance.

Significant Difference Between Two Means

1. Develop the null hypothesis or alternative hypothesis.

2. Find the difference of means between the treated group and the controlled group.

    $$\text{Diff} = \bar{x}_T - \bar{x}_C$$

3. Create a randomization distribution of the difference of means between two randomly divided groups A and B (the treated and controlled are mixed).

    $$\text{Diff}_1 = \bar{x}_A - \bar{x}_B$$
    $$\text{Diff}_2 = \bar{x}_A - \bar{x}_B$$
    $$\text{Diff}_3 = \bar{x}_A - \bar{x}_B$$
    • • • • • •

4. Compute the probability of getting these results.

5. Make a conclusion based on the significance level. Choose a significance level α that you regard as decisive. If the probability is less than or equal to α, you conclude that the alternative hypothesis is true; if it is greater than α, you conclude that the data does not provide sufficient evidence to reject the null hypothesis.

e.g. A math teacher believes that after school tutoring can help students improve their scores. What conclusion can we make based on the following data?

The scores of the 5 students with tutoring are:
91, 80, 77, 84, 63
The scores of the 5 students without tutoring are:
74, 75, 69, 62, 70

1. Alternative hypothesis: $\text{Diff} = \bar{x}_T - \bar{x}_C > 0$

2. Find the difference of means between the treated group and the controlled group.

    $$\text{Diff} = \bar{x}_T - \bar{x}_C = 79 - 70 = 9$$

3. Create a randomization distribution of the difference.

Method 1: Manual Simulation

Label these 10 numbers.
91 (1), 80 (2), 77 (3), 84 (4), 63 (5)
74 (6), 75 (7), 69 (8), 62 (9), 70 (10)

Use the graphing calculator to create a list of random numbers between 1 to 10.
[MATH] / PRB /5:randInt(1,10)[ENTER]• • •
Skip the repeated number in the same group.
Group A   Group B
(2, 7, 6, 3, 8)   (the rest of five numbers)
(7, 6, 2, 1, 5)
(5, 8, 10, 9, 2)
• • • • • •

Replace the labels by the real values:
Group A   Group B
(80, 75, 74, 77, 69)   (91, 84, 63, 62, 70)
(75, 74, 80, 91, 63)   (70, 77, 84, 69, 62)
(63, 69, 70, 62, 80)   (91, 77, 84, 74, 75)
• • • • • •

$$\text{Diff}_1 = \bar{x}_A - \bar{x}_B = 75 - 74 = 1$$
$$\text{Diff}_2 = \bar{x}_A - \bar{x}_B = 76.6 - 72.4 = 4.2$$
$$\text{Diff}_3 = \bar{x}_A - \bar{x}_B = 68.8 - 80.2 = -12.2$$
• • • • • •

Construct a dot plot or histogram.

Note: This method is conceptional, but not practical.

V. Probability and Statistics

Method 2: Computer Simulation

Use the application on the website:
www.rossmanchance.com/applets

Select: Statistical Inference
       Randomization test for quantitive response
       o two means
Clear the original sample data, enter the following:
    T   91
    T   80
    T   77
    T   84
    T   63
    C   74
    C   75
    C   69
    C   62
    C   70
Press [Use Data]
Select the check box next to "Show Shuffle Options"
Enter 500 in the Number of Shuffles box
Press [Shuffle Responses]

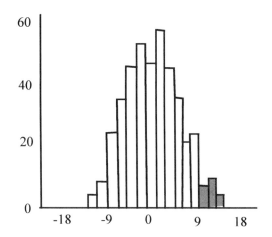

4. Compute the probability.

Since $\bar{x}_T - \bar{x}_C = 9$,
Count Samples [Greater Than ≥] [9]
Press [Count]

Count = 24/500 (0.0480)

Note: The results may vary slightly from time to time.

5. We can conclude that after school tutoring makes a significant difference at a 5% significance level.

29. STATISTICS (Bivariate Data)

Many statistics applications study the relationship between two variables, bivariate data.

Correlation: The relationship between two sets of data
Causation: The relationship in which one variable produces an effect on the other

There are two different types of data:
Quantitive Data is numerical in nature, such as age, height, size, temperature, distance, etc.
Qualitative Data is categorical in nature, such as gender, race, color, etc.

29.1 Exploring Categorical Data

A two-way frequency table is used to summarize bivariate categorical data.

e.g. Determine the favorite color of two groups of people: Male and Female

(1). A two-way frequency table:

	Red	Blue	Yellow	Total
Male	20	30	10	60
Female	20	15	5	40
Total	40	45	15	100

(2). A two-way relative frequency table:
    Each cell value is divided by the grand total.

	Red	Blue	Yellow	Total
Male	0.2	0.3	0.1	0.6
Female	0.2	0.15	0.05	0.4
Total	0.4	0.45	0.15	1

(3). A two-way conditional relative frequency table:
    Each cell value is divided by the marginal total.

	Red	Blue	Yellow	Total
Male	20/60	30/60	10/60	1
Female	20/40	15/40	5/40	1
Total	40/100	45/100	15/100	1

V. Probability and Statistics

29.2 Exploring Numerical Data

A scatter plot gives us the relationship between two numerical variables. Each point on the plot corresponds to a pair of ordered numbers (x, y).

A function model gives us an algebraic expression of the relationship between two numerical variables. The function gives the predicted value which is different from the actual value.

The actual value is denoted as Y.
The predicted value is denoted as y.

The prediction error is the difference between the actual value and the predicted value. This difference is called a **residual**.

residual $R = Y - y$

The best regression model should have the least amount of residuals.

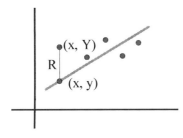

To measure the strength of the regression model, we use the **correlation coefficient r**, which is between -1 and 1. Positive values mean a positive relationship (increasing function), and negative values mean a negative relationship (decreasing function).
The strength of the relationship increases as r moves away from 0 toward either -1 or 1.

Residual Plots:

The residual Rs can be plotted against the x-values, which is called the **residual plot**.

If a pattern is seen in the residual plot, it indicates that a nonlinear model will be better for the relationship of the original data.

Use graphing calculator for residual plot.

[2nd] [STAT PLOT]
 1: plot 1 Off
 2: plot 2 On [ENTER]
   Xlist: L_1
   Ylist: RESID
      ([2nd][LIST]/NAMES/7: RESID)
[ZOOM] [9]

(1) Regression Modeling:

Linear Regression: $y = ax + b$

Correlation Coefficient r :

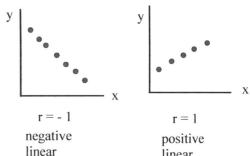

$r = -1$
negative linear relationship

$r = 1$
positive linear relationship

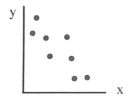

$r = -0.5$
moderate negative linear relationship

$r = 0.5$
moderate positive linear relationship

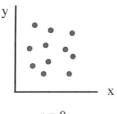

$r = 0$

no linear relationship

V. Probability and Statistics

(2) Line of Best Fit (The Least-Squares Line):

(1) Passing through the mean point $(\bar{x}, \bar{y})$.
(2) The difference between the model values and the real values is the least.

Use graphing calculator to find the equation of the Line of Best Fit : $y = ax + b$

e.g.

x_i	2	4	6	8	10
y_i	13	15	16	17	20

(1) Clear List L_1 and List L_2
[STAT] EDIT / 4: ClrList [ENTER] [2nd] [L_1]
[,] [2nd] [L_2] [ENTER]
(2) Enter data to L_1 and L_2
[STAT] EDIT / 1: Edit ... [ENTER]
Enter data x_i into List L_1 ; Enter data y_i into List L_2.
(3) Scatter Plot: [2nd] [STAT PLOT] 1: Plot 1
[ENTER]
ON
Type:

[ZOOM] [9]
(4) Find the equation of the Line of Best Fit and the Correlation Coefficient r :
[2nd] [CATALOG] Diagnostic On [ENTER]
[STAT] CALC / 4: LinReg(ax + b) [ENTER]
[2nd] [L_1] [,] [L_2] [ENTER]
LinReg $y = ax + b$
   $a = 0.8$ $b = 11.4$ $r = 0.98$

(5) Draw the Line of Best Fit:
[Y =] [VARS] 5: Statistics ... [ENTER] EQ / 1: RegEQ [ENTER] [ZOOM] [9]

(6) Predict the results by using the model:
Find the value of y when x = 10.5
[2nd] [CALC] 1: Values [ENTER]
X = 10.5 [ENTER]   Y = 19.8
(Adjust window dimensions for **extrapolation**, the process of finding values outside of the given range)

(3) Other Regressions:

5: QuadReg: $y = ax^2 + bx + c$
6: CubicReg: $y = ax^3 + bx^2 + cx + d$
9: LnReg: $y = a + b\ln x$
0: ExpReg: $y = a \cdot b^x$
A: PwrReg: $y = ax^b$

To find the best model for a given set of data, compare the values of r.
Correlation Coefficient $r = \pm 1$ means exact fit.

e.g.

x_i	0	1	2	3	4	5	6
y_i	5	10	20	40	80	160	320

(1) Enter the data into L_1 and L_2
(2) Make the scatter plot for the data
(3) Test the exponential model: $y = a \cdot b^x$
[STAT] CALC / 0: ExpReg [ENTER]
[2nd] [L_1] [,] [2nd] [L_2] [ENTER]
ExpReg $y = a \cdot b^x$ $a = 5$ $b = 2$ $r = 1$
   $y = 5 \cdot 2^x$
(4) Draw the graph of the model:
[Y =] [VARS] 5: Statistics ... [ENTER]
EQ / 1: RegEQ [ENTER] [ZOOM] [9]

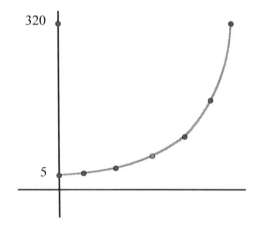

A1. Graphing Calculator

TI 84 Plus is used for the examples in this section.

1 Tips

Clear the Memory:
  [2nd] [MEM] 7: Reset ... [ENTER]
  1: All Ram ...[ENTER]   2: Reset [ENTER]
          Ram Cleared
  [2nd] [MEM] 7: Reset ... [ENTER]
  2: Defaults ...[ENTER]   2: Reset [ENTER]
          Defaults Set

Return to Home Screen:
  [2nd] [QUIT]

2 Graph Functions

e.g. Graph $y = -x^2 + 4$
  [y =] [(-)] [X, T, θ, n] [x^2] [+] [4] [GRAPH]

e.g. Graph $y = |x - 4|$
  [y =] [MATH] NUM / 1: abs [ENTER]
      [X, T, θ, n] [-] [4] [ )] [GRAPH]

Zoom Menu:
To have a better view of the graph:
  [ZOOM] 6: ZStandard (- 10 < x < 10 ; - 10 < y < 10)
  [ZOOM] 4: ZDecimal (to set Δx to 0.1)
  [ZOOM] 0: ZoomFit   (to see complete graphs)
  [ZOOM] 2: Zoom In   (to see details around cursor)
  [ZOOM] 7: ZTrig   (for Trigonometry ; $X_{scl} = \frac{\pi}{2}$)
  [ZOOM] 9: ZoomStat (for Statistics)

e.g. Graph $y = \sin 2x$
  [y =] [sin] [2] [X, T, θ, n] [ZOOM] [7]

Change Window Dimensions:
e.g. Graph $y = -3x^2 + 12x + 5$
  [y =] [(-)] [3] [X, T, θ, n] [x^2] [+] [1] [2]
  [X, T, θ, n] [+] [5] [ZOOM] [6]

  To see the complete graph:
  [WINDOW] $Y_{max} = 20$ [ENTER] [GRAPH]

Trace:
  To see how y values vary with x values:
    [TRACE]
  To find the value of y at a specific value of x:
    [2nd] [CALC] 1: value [ENTER]

3 Table of a Function

e.g. Display the table of $y = x^2$
  [y =] [X, T, θ, n] [x^2]
  [2nd] [TABLE]
  To change the x increment:
  [2nd] [TBLSET] ΔTbl

4 Solve Equations

e.g. Solve $x^2 - 9 = 0$
  (1) Graph $y = x^2 - 9$
  (2) [2nd] [CALC] 2: zero [ENTER]
  (3) Move the cursor to set the Left Bound
      [ENTER] and the Right Bound [ENTER]
      of the x - intercept, then Guess [ENTER]
          Zero x = - 3   y = 0
  (4) Repeat (3) to find the other zero
          Zero x = 3   y = 0

5 Solve the System of Equations

e.g. Solve system $xy = 8$
              $y = x + 2$
  (1) Rewrite the first equation as $y = \frac{8}{x}$
  (2) Graph those two functions
  (3) [2nd] [CALC] 5: intersect [ENTER]
  (4) $Y_1 = 8/X$
      First Curve?
  Move the cursor to the intersection [ENTER]
      $Y_2 = X + 2$
      Second Curve ? [ENTER]
      Guess? [ENTER]
    Intersection X = 2   Y = 4
  (5) Repeat (4) to find the other intersection:
          X = -4   Y = -2

6 Maximum and Minimum

e.g. Find the maximum or minimum of the function
      $y = x^2 - 6x + 3$
  (1) Graph $y = x^2 - 6x + 3$
  (2) [2nd] [CALC] 3: minimum [ENTER]
  (3) Move the cursor to set the Left Bound
      [ENTER] and the Right Bound [ENTER]
      of the minimum, then Guess [ENTER]
          Minimum x = 3   y = -6
      when x = 3 the function has a minimum of -6 .

A2. Algebraic Formulas

Polynomials:
$a^2 - b^2 = (a - b)(a + b)$
$a^3 - b^3 = (a - b)(a^2 + ab + b^2)$
$a^3 + b^3 = (a + b)(a^2 - ab + b^2)$
$a^n - b^n = (a - b)(a^{n-1} + a^{n-2}b + a^{n-3}b^2 \ldots\ldots + ab^{n-2} + b^{n-1})$

Binomial Expansions:
$(a + b)^2 = a^2 + 2ab + b^2$
$(a - b)^2 = a^2 - 2ab + b^2$
$(a + b)^3 = a^3 + 3a^2b + 3ab^2 + b^3$
$(a - b)^3 = a^3 - 3a^2b + 3ab^2 - b^3$
$(x + y)^n = {_nC_0}x^n y^0 + {_nC_1}x^{n-1}y^1 + {_nC_2}x^{n-2}y^2 \ldots\ldots {_nC_{n-1}}x^1 y^{n-1} + {_nC_n}x^0 y^n$

Combination: $\quad {_nC_r} = \dfrac{{_nP_r}}{r!} \qquad (r \leq n)$

Permutation: $\quad {_nP_r} = n(n - 1)(n - 2)(n - 3) \bullet\bullet\bullet\bullet\bullet\bullet (n - r + 1) \qquad$ (r factors)

Factorial: $\quad r! = r(r - 1)(r - 2)(r - 3) \bullet\bullet\bullet\bullet\bullet\bullet 3 \bullet 2 \bullet 1$

Completing the Square: $x^2 + bx + (\dfrac{b}{2})^2 = (x + \dfrac{b}{2})^2$

Quadratic Formula: $ax^2 + bx + c = 0 \qquad$ where $a \neq 0$

$$x = \dfrac{-b \pm \sqrt{b^2 - 4ac}}{2a}$$

Laws of Exponents:

$a^m a^n = a^{m+n}$, $\quad (a^m)^n = a^{mn}$, $\quad a^{\frac{m}{n}} = \sqrt[n]{a^m}$, $\quad (ab)^n = a^n b^n$

$\dfrac{a^m}{a^n} = a^{m-n}$, $\quad a^{-n} = \dfrac{1}{a^n}$, $\quad a^0 = 1$

Laws of Logarithms:

$\log_a AB = \log_a A + \log_a B$, $\quad \log_a A^n = n \bullet \log_a A$, $\quad \log_a \sqrt[n]{A} = \dfrac{1}{n}\log_a A$

$\log_a \dfrac{A}{B} = \log_a A - \log_a B$, $\quad \log_a 1 = 0$, $\quad \log_a a = 1$

Change of Base Formula:

$\log_a A = \dfrac{\log A}{\log a}$, $\quad \log_a A = \dfrac{\ln A}{\ln a}$, $\quad a^x = e^{x \ln a}$

Absolute Values: $|x| = \sqrt{x^2}$, $\quad |ab| = |a||b|$, $\quad |a + b| \leq |a| + |b|$

Index

A.A. similar, 20
A.A.S. congruent, 19
A.S.A. congruent, 19
Absolute error, 36
Absolute maximum, 40
Absolute minimum, 40
Absolute value equations, 6
Absolute value function, 45
Absolute value inequality, 7
Acute angle, 16
Addition of complex numbers, 58
Addition of polynomials, 3
Addition or subtration method, 11
Addition property, 15
Additive identity, 2
Additive inverse, 2
Algebraic expressions, 5
Alternate interior angles, 16
Alternative hypothesis, 83
Altitude, 17
Ambiguous case, 72
Amplitude, 69
Analyze data, 76
Angle bisector theorem, 20
Angle bisector, 17
Angle bisector, equation of, 63
Angle measure, 23
Angle of depression, 67
Angle of elevation, 67
Angles of a circle, 23
Angles, 16
Annual percentage rate, 50
Arc length, 26, 65
Arc, 23
Area formula of a triangle, 72
Area of sector, 26
Areas, 34, 35
Arithmetic means, 12
Arithmetic sequences, 12
Arithmetic series, 13
Associative property, 2
Assumption of sampling distributions, 81
Asymptote, 49
Average rate of change, 40
Average speed, 40
Average velocity, 40

Axis of symmetry, 46
Base area, 34, 35
Basic algebraic formulas, 88
Benchmark, 76
Bernoulli experiment, 75
Bias, 76
Biconditional statements, 15
Binomial expansions, 75
Binomial probability, 75
Bisector of a segment, 16
Bisector of an angle, 16
Bivariate data, 83
Box plot, 76
Categorical data, 84
Causation, 84
Cause and effect conclusions, 76
Cavalieri's principle, 33
Center of gravity, 17
Center-radius equation of a circle, 63
Central angle, 23
Central tendency of the data, 76
Centroid, 17, 37
Certain case, 74
Change of base formula, 53
Chord - chord angle, 24
Chord - tangent angle, 24
Chord, 23
Circle measurements, 34
Circle, 23
Circle, equation of, 48
Circumcenter, 17
Circumference, 34
Circumscribed circle, 17
Clear the memory, 87
Cofunctions, 66
Collecting data, 76
Combination, 73
Combine like terms, 3
Common logarithms, 52
Common tangents of two circles, 26
Commutative property, 2
Complement of a set, 1
Complementary, 16
Complete the square, 7, 48
Complex numbers, 58
Composition of functions, 41

Index

Composition of transformations, 30
Compound inequalities, 6
Compound loci, 28
Computer simulation, 83
Concurrence, 17
Conditional statements, 15
Cone, 35
Confidence interval, 81
Congruence, 31
Congruent angles, 16
Congruent arcs, 26
Congruent segments, 16
Congruent triangles, 19
Conic sections, 54
Conjugates, 4, 59
Conjunction, 15
Constant function, 39
Constructions, 27
Contrapositive statements, 15
Controlled experiments, 76
Converse statements, 15
Coordinate geometric proof, 37
Coordinate plane, 37
Coplanar, 32
Correlation coefficient, 85
Correlation, 84
Corresponding angles, 16
Coterminal angles, 65
Counting principle for probability, 75
Counting principle, 73
CPCTC, 19
Cross-multiplication, 9
Cube measurements, 34
Cumulative frequency histogram, 77
Cumulative frequency table, 77
Cyclic quadrilateral, 26
Cylinder, 35
Data distribution, 78
Decreasing function, 39
Definitions in geometry, 16
Degree measure, 65
Density curve, 79
Diagonals, 22
Difference of two squares, 3
Dihedral angle, 32
Dilation, 29, 42

Direct isometry, 31
Direct variation, 43
Directrix, 54
Discriminant of the quadratic equation, 8
Disjunction, 15
Dispersion of the data, 76
Distance between two complex numbers, 60
Distance from a point to a line, 44
Distance, 37
Distributive property, 2
Dividend, 55
Division in polar form, 61
Division of complex numbers, 60
Division of polynomials, 3
Division property, 15
Divisor, 55
Domain, 38
Dot plot, 76
Double angle formulas, 71
e, the number, 14
Ellipses, 54
End behavior, 56
Equations, 6
Equidistant, 17
Equilateral triangles, 17, 18
Error in measurement, 36
Even function, 40
Event, 74
Exponential decay, 49
Exponential equations, 49
Exponential functions, 49
Exponential growth, 49
Exponential inequalities, 50
Exponential models, 49
Exponents, 5
Exterior angles, 17, 22
Extrapolation, 86
Factor theorem, 56
Factored form of quadratic function, 46
Factorial, 73
Factoring polynomials, 3, 61
Fibonacci sequence, 12
Finite sequence, 11
Finite series, 13
First degree function, 43
Five-number summary, 76

Index

Foci, 54
Focus, 54
Force problems, 72
Formulas for measuring, 34
Frequency histogram, 77
Frequency table, 77
Frequency, 69, 76
Frustum of right circular cone, 35
Frustum of sphere, 36
Function notation, 39
Function table, 87
Functions under a transformation, 42
Functions, 38
Fundamental loci, 28
Fundamental theorem of algebra, 62
Future value, 50
General form of linear function, 43
General form of quadratic function, 46
Geometric means, 12
Geometric measurements, 33
Geometric sequences, 12
Geometric series, 14
Glide reflection, 30
Graph functions by graphing calculator, 87
Graphing calculator, 87
Greatest common factor (GCF), 3
Greatest integer function, 45
H.L. congruent, 19
Half angle formulas, 71
Half-life, 49
Home screen, 87
Horizontal asymptote, 49
Horizontal line test, 38
Horizontal lines, 44
Hyperbolas, 54
Hypotenuse, 18
Identity symmetry, 31
Imaginary number i, 58
Impossible case, 74
Incenter, 17
Increasing function, 39
Index of sigma notation, 13
Indirect proof, 21
Inequalities, 6
Infinite sequence, 11
Infinite series, 13, 14

Infinities, 56
Inscribed angle theorem, 23
Inscribed angle, 23
Inscribed circle, 17
Inscribed polygon, 26
Intercept form, 43
Interest compounding, 50
Interior angles, 17, 22
Interquartile range (IQR), 76
Intersection of two sets, 1
Interval notation, 1, 39
Inverse functions, 41
Inverse statements, 15
Inverse trigonometric functions, 70
Inverse variations, 43
Isometry, 31
Isosceles trapezoid, 22
Isosceles triangles, 17, 18
Lateral area, 34, 35, 36
Lateral edges, 34
Law of cosines, 72
Law of sines, 72
Least Common Denominator (LCD), 9
Least-squares line, 86
Legs, 18
Like terms, 3
Line of best fit, 86
Line reflection, 29
Line symmetry, 31
Linear equations, 6
Linear factorization theorem, 62
Linear function, 43
Linear inequalities, 6, 45
Linear system of equations, 11
Linear system, graphic solutions of, 64
Local maximum, 40
Local minimum, 40
Loci, equations of , 63
Locus, loci, 28
Logarithmic equations, 53
Logarithmic functions, 52
Logarithmic inequalities, 53
Logic, 15
Long division algorithm, 55
Lower limit of sigma notation, 13
Lower quartile, 76

Index

Major axis, 54
Map, 29
Margin of error, 81
Maximum by graphing calculator, 87
Maximum height, 47
Maximum, 40
Mean, 76
Median of a trapezoid, 22
Median, 17, 76
Middle quartile, 76
Midline, 69
Midpoint of two complex numbers, 60
Midpoint, 16, 37
Midsegment theorem, 21
Minimum by graphing calculator, 87
Minimum, 40
Minor axis, 54
Mode, 76
Modulus of a complex number, 58, 60
Mortgage loan and payments, 50
Multiplication in polar form, 61
Multiplication of complex numbers, 59
Multiplication of polynomials, 3
Multiplication property, 15
Multiplicative identity, 2
Multiplicative inverse, 2, 59
Multiplicity of zeros, 56
Natural logarithms, 52
Negation, 15, 74
Newton's law of cooling, 51
Nonadjacent interior angles, 17
Noncollinear points, 32
Normal curve, 79
Normal distribution, 79
Null hypothesis, 83
Numbers, 1
Numerical data, 85
Oblique cylinder, 33
Observations, 76
Obtuse angle, 16
Odd function, 40
One-to-one function, 38
Opposite isometry, 31
Ordered pair, 37
Orientation, 31
Origin, 37

Orthocenter, 17
Outliers, 76
Parabolas, 46, 54
Parallel lines, 16
Parallel planes, 33
Parallelogram measurements, 34
Parallelogram, 22
Partial sum, 13
Partition postulate, 15
Percent of error, 36
Percentage, 2
Percentile, 76
Period, 69
Permutation, 73
Perpendicular bisector, equation of, 63
Perpendicular bisectors, 17
Perpendicular lines, 16
Perpendicular planes, 32
Phase shift, 69
Piecewise defined function, 39
Plane, 32
Point reflection, 29
Point symmetry, 31
Point-slope form, 43
Polar axis, 60
Polar form of a complex number, 60
Polygons, 22
Polynomial division, 55
Polynomial expressions, 55
Polynomial formulas, 55
Polynomial functions, 55
Population characteristics, 76, 81
Population parameter, 81
Population, 76
Postulates of equality, 15
Postulates of inequality, 15
Postulates, 15
Powers in polar form, 61
Present value, 50
Principal, 50
Prism measurements, 34
Probability of normal distribution, 80
Probability, 73
Projectile motion, 47
Proofs, 17, 19, 37
Proportion, 2, 9, 20

Index

Ptolemy's theorem, 26
Pyramid, 35
Pythagorean theorem, 18
Pythagorean triples, 18, 66
Quadrantal angles, 66
Quadrants, 37
Quadratal angles, 66
Quadratic equations, 7, 61
Quadratic formula, 7, 61
Quadratic function, 46
Quadratic inequalities, 8, 47
Quadratic-linear system, 11, 64
Quadrilateral, 22
Qualitative data, 84
Quartile, 76
Quotient, 55
Radian measure, 65
Radical equations, 10
Radical expressions, 4
Radical operations, 4
Randomization distribution, 83
Range, 38, 76
Rate of change, 43
Ratio, 2, 20
Rational equations, 9
Rational exponents, 5
Rational expressions, 4
Rational inequalities, 10
Rational operations, 4
Rational zeros theorem, 57
Rationalization of denominator, 4
Rearrange the formula, 5
Reciprocal trigonometric functions, 65
Reciprocal, 60
Rectangle measurements, 34
Rectangle, 22
Rectangular form of a complex number, 60
Rectangular prism, 34
Rectangular region, 45
Recursive definition, 12
Reference angle, 67
Reflection, 29, 42
Reflexive property, 15
Regression models, 85, 86
Regular polygon, 22
Regular pyramid, 35

Relations, 38
Relative error, 36
Relative frequency histogram, 77
Relative frequency table, 77
Relative maximum, 40
Relative minimum, 40
Remainder theorem, 56
Remainder, 55
Repeated zeros, 56
Residual plots, 85
Restricted domain, 38, 41
Resultant vector, 58
Rhombus measurements, 34
Rhombus, 22
Right angle, 16
Right circular cone, 35
Right circular cylinder, 33, 35
Right prism, 34
Right triangle altitude theorem, 21
Right triangle, 18
Rigid motion, 31
Roots in polar form, 61
Roots of the function, 56
Roots of the quadratic equation, 8
Rotation, 29
Rotational symmetry, 31
S.A.S. congruent, 19
S.A.S. similar, 20
S.S.S. congruent, 19
S.S.S. similar, 20
Sample space, 73, 74
Sample statistic, 81
Sample survey, 78
Sample, 76
Sampling variability, 81
Scale drawing, 29
Scale factor, 29
Scalene triangles, 17
Scatter plot, 85
Scientific notation, 5
Secant - secant angle, 24
Secant line, 23
Segments of a circle, 25
Semicircle, 23
Sequence, 11
Series, 13

Index

Set notation, 1, 39
Set, 1
Sigma notation, 13
Significance level, 83
Signs of trigonometric functions, 66
Similar objects, 33
Similar triangles, 20
Similarity, 31
Simple random sample (SRS), 81
Simulation, 83
Sketch polynomial functions, 57
Skew lines, 32
Slant height, 35
Slope, 37
Slope-intercept form, 43
Slopes of lines, 44
Solid geometry, 32
Solve equations by graphing calculator, 87
Sphere, 36
Square measurements, 34
Square, 22
SRS, (simple random sample), 81
Standard deviation, 78
Standard error, 81
Standard form of a polynomial, 55
Statistical inference, 81
Statistics, 76
Substitution method, 11
Substitution property, 15
Subtraction of complex numbers, 59
Subtraction of polynomials, 3
Sum and difference formulas, 70
Supplementary, 16
Surface area, 34, 35, 36
Surveys, 76
Symmetric property, 15, 20
Symmetry, 31
System of inequalities, 64
Tangent - secant angle, 24
Tangent - tangent angle, 24
Tangent line, 23
Tangent segment, 24
Tax problems, 2
Thales' theorem, 23
Theorems, 16
Transformations, 29

Transitive property, 15, 20
Translation, 29, 42
Transversal, 16
Trapezoid measurements, 34
Trapezoid, 22
Tree diagram, 73
Triangle inequalities, 17
Triangle measurements, 34
Triangles, 17
Triangular region, 45
Trigonometric applications, 67, 72
Trigonometric equations, 71
Trigonometric functions, 65
Trigonometric graphs, 68, 69
Trigonometric identities, 70
Trigonometric identity proof, 70
Turning point, 46
Union of two regions, 33
Union of two sets, 1
Unit circle, 66
Univariate data, 76
Universal set, 1
Upper limit of sigma notation, 13
Upper quartile, 76
Variance, 78
Venn diagram, 73
Vertex form of quadratic function, 46
Vertex, 17
Vertical angles, 16
Vertical launch, 47
Vertical line test, 38
Vertical lines, 44
Volume, 34, 35, 36
Window dimensions, change, 87
X-axis, 37
X-intercept, 56
Y-axis, 37
Zero of the function, 56
Zero with multiplicity, 56
Zoom menu, 87
Z-score, 78

Also Available

Student's Choice
Regents Review Integrated Algebra   ISBN: 9781453880982

Student's Choice
Regents Review Geometry   ISBN: 9781453709993

Student's Choice
Regents Review Algebra 2/Trigonometry   ISBN: 9781460983874

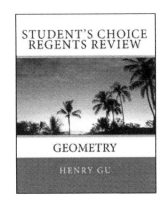

The Complete Calculus Review Book   ISBN: 9781478362203

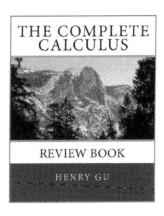

Made in the USA
San Bernardino, CA
16 January 2016